Lecture Notes in Computer Science 7075

Commenced Publication in 1973
Founding and Former Series Editors:
Gerhard Goos, Juris Hartmanis, and Jan v

Henning Müller   Hayit Greenspan
Tanveer Syeda-Mahmood (Eds.)

Medical Content-Based Retrieval for Clinical Decision Support

Second MICCAI International Workshop, MCBR-CDS 2011
Toronto, ON, Canada, September 22, 2011
Revised Selected Papers

 Springer

Volume Editors

Henning Müller
University of Applied Sciences
Western Switzerland
TechnoArk 3
3960 Sierre, Switzerland
E-mail: henning.mueller@hevs.ch

Hayit Greenspan
Tel Aviv University
Department of Biomedical Engineering
The Iby and Aladar Fleischman Faculty of Engineering
Ramat Aviv, Israel
E-mail: hayit@eng.tau.ac.il

Tanveer Syeda-Mahmood
Multi-modal Mining for Healthcare
IBM Almaden Research Center
650 Harry Road, San Jose, CA 95120, USA
E-mail: stf@almaden.ibm.com

ISSN 0302-9743   e-ISSN 1611-3349
ISBN 978-3-642-28459-5   e-ISBN 978-3-642-28460-1
DOI 10.1007/978-3-642-28460-1
Springer Heidelberg Dordrecht London New York

Library of Congress Control Number: 2012931414

CR Subject Classification (1998): J.3, I.5, H.2.8, I.4, H.3, H.5

LNCS Sublibrary: SL 6 – Image Processing, Computer Vision, Pattern Recognition, and Graphics

Typesetting: Camera-ready by author, data conversion by Scientific Publishing Services, Chennai, India

Printed on acid-free paper

Springer is part of Springer Science+Business Media (www.springer.com)

Preface

This document contains articles from the Second Workshop on Medical Content-Based Retrieval for Clinical Decision Support (MCBR-CDS) that took place at the MICCAI (Medical Image Computing for Computer-Assisted Intervention) 2011 conference in Toronto, Canada, during September 22, 2011. The first workshop on this topic took place at MICCAI 2009 in London, UK. An earlier workshop on medical image retrieval was conducted at MICCAI 2007, Brisbane, Australia

The workshop obtained 17 high-quality submissions of which 11 were selected for presentation after three external reviewers and one workshop organizer reviewed each of the papers. The review process was double blind.

Although this was only a small workshop, the quality of the submissions compared to the 2009 workshop had significantly increased and thus also the acceptance rate of above 50% was justified. The papers were from a total of nine different countries and from four continents, highlighting the diversity of submissions.

At the workshop two invited presentations were given in addition to the 11 oral presentations. Two invited speakers gave overviews from state-of-the-art academic research (Nicholas Ayache, INRIA, France) and industrial (Dorin Comaniciu, Siemens Corporate Research, USA) perspectives in the domain. A panel at the end discussed the role of content-based image retrieval in clinical decision support. In general, the workshop resulted in many lively discussions and showed well the current trends and tendencies in content-based medical retrieval and how this can support decisions in clinical work.

These proceedings contain the 11 accepted papers of the workshop as well as the invited presentation given by Nicholas Ayache on image retrieval.

An overview of the workshop is included first, summarizing the papers and the discussions that took place at the workshop itself.

We would like to thank all the reviewers that helped make a selection of high-quality papers for the workshop. The many comments also made the presented papers much better than their initial versions. We hope to have a similar workshop in next year's MICCAI conference.

September 2011

Henning Müller
Hayit Greenspan
Tanveer Syeda-Mahmood

Organization

Organizing Committee

General Co-chairs   Hayit Greenspan, Israel
Henning Müller, Switzerland
Tanveer Syeda-Mahmood, USA

Publication Chair   Henning Müller, Switzerland

International Program Committee

Burak Acar	Bogazici University, Turkey
Amir Amini	University of Louisville, USA
Sameer Anatani	National Library of Medicine (NLM), USA
Rahul Bhotika	GE Global Research Center, NY, USA
Albert Chung	Hong Kong University of Science and Technology, Hong Kong
Antonio Criminisi	Microsoft Research, Cambridge, UK
Thomas M. Deserno	Aachen University of Technology (RWTH), Germany
Gerhard Engelbrecht	University Pompeu Fabra (UPF), Spain
Bram van Ginneken	Radboud University Nijmegen Medical Centre, The Netherlands
Allan Hanbury	Vienna University of Technology, Austria
Nico Karssemeijer	Radboud University Nijmegen, The Netherlands
Jayashree Kalpathy-Cramer	Harvard University, USA
Georg Langs	MIT, USA
Yanxi Liu	UPENN, USA
Rodney Long	NLM, USA
Robert Lundstrum	Kaiser Permanente, San-Francisco Medical center, USA
Kazunori Okada	SFSU, USA,
Daniel Racoceanu	French National Center for Scientific Research (CNRS), France
Daniel Rubin	Stanford, USA
Linda Shapiro	University of Washington, USA
Ron Summers	NIH, USA
Agma Traina	University of Sao Paulo, Brazil
Pingkun Yan	Chinese Academy of Sciences, China
S. Kevin Zhou	Siemens Corporate Research, USA

Sponsors

European Commission 7th Framework Programme, projects Khresmoi (257528), Promise (Promise), Chorus+ (249008) and the Swiss National Science Foundation (205321–130046). Thanks also to IBM for their support.

Table of Contents

Applications

Multidimensional Retrieval

Overview of the Second Workshop on Medical Content–Based Retrieval for Clinical Decision Support

Adrien Depeursinge[1], Hayit Greenspan[2], Tanveer Syeda–Mahmood[3],
and Henning Müller[1]

[1] University of Applied Sciences Western Switzerland, Sierre, Switzerland
[2] Tel Aviv University, Israel
[3] IBM Almaden Research Lab, USA
henning.mueller@hevs.ch

Abstract. The second workshop on Medical Content–Based Retrieval for Clinical Decision Support took place at the MICCAI conference in Toronto, Canada on September 22, 2011. The workshop brought together more than 40 registered researchers interested in the field of medical content–based retrieval. Eleven papers were accepted and presented at the workshop. Two invited speakers gave overviews on state–of–the–art academic research and industrial perspectives. The program was completed with a panel discussion on the role of content–based retrieval in clinical decision support. This overview introduces the main highlights and discussions in the workshop, summarizes the novelties and introduces the presented papers, which are provided in these proceedings.

1 Introduction

Medical content–based retrieval has received a large amount of research attention over the past 15 years [1, 2]. Despite large amounts of research and also the availability of benchmarking data sets for medical retrieval [3], there are still only few tests of real systems in clinical practice such as [4]. Image retrieval has always been at the crossing between several sciences, namely image analysis or computer vision, information retrieval and medical informatics. The MICCAI (Medical Image Computing for Computer–Assisted Intervention) conference is a logical target for a workshop on content–based retrieval.

A workshop titled *Content–based Image Retrieval for Biomedical Image Archives: achievements, problems and prospects* focusing on medical image retrieval, took already place in MICCAI 2007 in Brisbane, Australia. The first workshop on *Medical Content–based Retrieval for Clinical Decision Support*, took place at MICCAI 2009 in London, United Kingdom [5]. The motivation was to show that in addressing true clinical challenges, purely visual image retrieval needs to be augmented with additional information, such as text (coming from various sources including image captions or the patient record). Thus, the challenges in content–based retrieval for clinical applications are in fact multimodal

H. Müller et al. (Eds.): MCBR-CDS 2011, LNCS 7075, pp. 1–11, 2012.

in nature and require the combination of state–of–the–art tools for each modality and the fusion across the modalities to obtain optimal results.

The second workshop on Medical Content–Based Retrieval for Clinical Decision Support took place at the MICCAI conference in Toronto, Canada on September 22, 2011. The workshop brought together more than 40 registered researchers. The 2011 workshop web page was set up at[1] to advertise the event. Submissions were requested in the following principle areas of interest:

- data mining of multimodal medical data,
- machine learning of disease correlations from mining multimodal data,
- algorithms for indexing and retrieval of data from multimodal medical databases,
- disease model–building and clinical decision support systems based on multimodal analysis,
- practical applications of clinical decision support using multimodal data retrieval or analysis,
- algorithms for medical image retrieval or classification using the ImageCLEF collection.

A specific goal was also to promote the data sets created in the ImageCLEF[2] challenge [6–8]. Using standard data sets can help identifying well–performing techniques and help measuring performance improvements within and across techniques and systems. Several articles in the workshop used standard data sets, which allows to well judge the obtained performance.

In total, 17 papers were submitted to the workshop. All papers were reviewed by at least three external reviewers from the scientific committee as well as one member from the organization committee. Through this process 11 high-quality papers were selected for oral presentation at the workshop. In addition to the scientific papers, two invited speakers gave insights into their current research directions and projects. One invited speaker was from the academic field (Nicholas Ayache, INRIA, France) and one from industry (Dorin Comaniciu, Siemens research, USA). The workshop finished with a panel that discussed the role of medical content–based retrieval for clinical decision support.

The workshop led to many lively discussions on application areas, technologies (particularly the use of various types of visual words) and future ideas for the current tools. Feedback from the participants was very positive for continuing the workshop series at future MICCAI conferences.

2   Highlights of the Presentations

This section details the main highlights of the workshop by discussing the main novelties presented in the invited presentations and the scientific papers. Papers on computer–aided diagnosis for a specific medical application are presented in Section 2.2 and general management of visual information in large databases in Section 2.3.

---

[1] http://www.mcbr-cds.org/

[2] http://www.imageclef.org/

2.1 Main Novelties Presented

The workshop discussed several novel ideas in the field of medical content–based retrieval. Dominant was the application of many types of *visual words* and bag of visual words (BoVW) techniques [9–14]. Key to strong performance is the combination of feature space definition and strong machine learning tools. The optimization of BoVW remains an interesting research topic as the parameters can vary widely depending on the type of application.

The use of 3D data was introduced in the current workshop in several of the works presented [10, 15, 13, 16, 17]. This contrasts with past years, where most approaches were concentrating on retrieval in 2D images. Visual words were shown to be useful in the 3D space.

A key feature in the use of visual words is the need to define interest points or *salient points* [9, 13, 17]. Various techniques can be used — from random sampling to dense sampling (and others). The technique to give best performance needs to be tested for each application. For example, it has been shown that dense sampling is optimal if a specific subset of images is to be analyzed such as lung CTs or radiographs of various anatomic regions. This has been shown to give best results in the ImageCLEF 2009 competition [18], as well as in several of the works presented in the current workshop.

Another presented challenge is the combination of *text and visual retrieval* [15, 12, 19]. *Cross–modality retrieval* is presented in [16]. A less frequently studied subject is the *perceived visual similarity* [9]. This domain is of importance for applications that aim at the integration into clinical practice.

Several papers dealt with *efficiency* questions using a variety of approaches such as *thumbnails* [17], *inverted files* [16] and *random clustering* [14]).

In summary, the works presented covered a large variety of techniques and applications, addressing state–of–the–art topics in medical content–based retrieval. Initial work with defined clinical applications, as well as evaluation methodologies, were presented. Both are of great importance for future acceptance in the medical community.

In the years ahead, we can expect to see medical image retrieval applications integrated into viewing stations and into clinical information systems, as complementary to text analysis tools.

2.2 Computer–Aided Diagnosis

In [9], André et al. introduce a smart atlas for videos from probe–based confocal laser endomicroscopy (pCLE) using CBIR. The videos are transformed to 2D mosaic images built from the course of the probe [20]. pCLE mosaic images are a good candidate for CBIR, because little experience exists with this relatively novel imaging modality. Therefore, retrieving similar images with attached diagnosis can support both training and diagnosis. The bag–of–visual–words method [21] is used to extract prevailing visual concepts in a feature space spanned by the scale invariant feature transform (SIFT) descriptors [22] using a dense sampling. The similarity between pCLE videos is computed as the χ^2

distance between the histograms of the occurrences of the various visual words (VW) over the whole image. The evaluation of the retrieval performance addresses an important point of CBIR systems for clinical decision support, where retrieving visually similar images with distinct diagnoses provides important cues to assist the decision of the clinicians. Indirect and direct evaluations of the system are performed: the indirect performance is computed as the classification performance based on the pathological classes (e.g., benign, neoplastic), and a direct retrieval performance based on the perceived visual similarity. The latter, which may closely assess the needs of the clinicians, is based on a visual similarity score obtained from endoscopists using an online survey tool. Such acquired ground truth is used to estimate the interpretation difficulty as well as to learn the perceived similarity by "shortening" the distance between instances perceived as very similar. The consistency between VWs and eight visual semantic concepts defined by experts in pCLE is also verified.

Local 3D texture quantification is used by Burner et al. [10] to retrieve high–resolution computed tomography (HRCT) images of the lungs with emphysema and metastases. The system uses a region of interest (ROI) delineated by the user as a query. It then searches for HRCT images containing regions with similar texture properties. The regional texture properties are encoded using bags of VW in a feature space spanned by a multiscale 3D extension of local binary patterns (LBP) [23]. In the database, the ROIs are defined using the supervoxel algorithm [24] to divide the lung parenchyma into homogeneous regions. The distance between ROIs are computed using the diffusion distance [25] between the histograms of VW. The evaluation shows that the proposed method outperforms approaches based on a global similarity measure.

A system for the retrieval of similar ROIs in HRCT images from patients affected by interstitial lung diseases is proposed by Foncubierta et al. in [11]. VWs are used to encode the characteristics of 6 types of lung tissue expressed in a wavelet domain. The influences of two intrinsic parameters of the proposed methods are investigated. First, the importance of scaling parameters of difference of Gaussians (DoG) is studied by varying the number of supplementary intermediate scales when compared to the classical dyadic wavelet transform. The authors show that the scale progression does not have a significant influence on the retrieval performance, since the classical dyadic scheme allows the best performance. The optimal size of the visual vocabulary is also investigated. Results show that for classes with high intra–class variations such as healthy tissue, a high number of VWs is required. As soon as a sufficient number of VWs is reached, the performance remains stable with a slight decrease that can be explained as an effect of the curse of dimensionality.

An image retrieval system based both on visual and text attributes is proposed by Costa et al. in [15] to assist the diagnosis of hepatic lesions in CT. The system uses ROIs defined by the user as queries. The visual attributes consist of grey–level distributions and moments of the Hounsfield units in the ROIs as well as in the whole liver. Text attributes consist of 20 labels defined by clinicians to be relevant for the characterization of liver lesions in CT. The labels are attached to each lesion by two clinical experts. Intrinsic random forests [26] are used to assess the similarity between the lesions by counting the number of times that two instances appear in the same leaves. The proposed approach allows retrieval of similar lesions with high semantics and efficiency.

In [27], Safi et al. propose a computer-aided diagnosis system for the classification of pigmented skin dermoscopic images. In a first step, the global images are segmented into three regions indicating 1) healthy skin, 2) bright parts of the melanoma, and 3) dark parts of the melanoma using energy constraints in the CIE color space representation [28]. Then, visual features are extracted from the region corresponding to the dark parts of the melanoma. These consist of shape, color and texture properties as well as geometric attributes defined as important by the clinicians. A support vector machine (SVM) classifier with a polynomial kernel is used to classify skin regions in a compact feature representation obtained with the prevailing dimensions of principal component analysis (PCA). The methods are achieving an overall good performance on a dataset of 4240 benign and 232 malignant images.

2.3 Visual Data Management

An image retrieval system based both on visual information and text is used in [12] for the management of a large collection of histological images using the query by example paradigm. Visual content is represented using 500 VWs extracted from densely sampled patches expressed in terms of their discrete cosine transform of the three RGB channels. The occurrences of the VWs are expressed using a matrix $X_v \in \mathbb{R}^{n \times l}$, where n is the number of VWs and l is the number of instances. Text attributes are obtained from expert annotations and are represented as 46 binary attributes that form the matrix $X_t \in \mathbb{R}^{m \times l}$. Reductions of the dimensionality of the feature space are obtained by factorizing the matrices X_v and X_t. A distance based on the scalar product of the vectors with reduced dimensionality is used for retrieval. Several approaches are proposed and compared to obtain a multimodal representation of the images. The first one consists of creating a multimodal matrix from the concatenation of X_v and X_t. The second maps X_v on the factorized representation H of X_t using a linear transform such as $X_v = W_v H$. W_v i s obtained using multiplicative updating rules [29]. A third strategy consists of mapping X_v on X_t as $X_v = W X_t$ directly with $W \geq 0$. A baseline using the histogram intersection of the occurrence of VWs as similarity metric is shown to be outperformed by all multimodal retrieval approaches but the one based on the concatenation of the visual and textual attributes.

The combination of textual and visual attributes for image retrieval is also investigated by Rahman et al. in [19]. The ImageCLEF 2010 dataset of the medical image retrieval task [30] is used, in which each image has text attached in the form of an image caption and text from the scientific journals from which the image belongs. The text attached to images is represented using MeSH terms[3] resulting from a preprocessing step consisting of the removal of stop words. The

---

[3] http://www.nlm.nih.gov/mesh/

vector of text attributes is obtained with the vector space model [31] and the importance of each term is weighted using the $tf - idf$ scheme with local (i.e., at the document level) and global (i.e., document collection) weights. Visual attributes are obtained from an SVM–classification of image patches using color and texture information into a set of 30 visual concepts defined by annotations from experts. The similarity between images is computed as a linear combination of two cosine distance measures from each modality. The results returned for a given query are filtered by modality using a previously trained SVM classifier. The evaluation shows that the combination of text and visual features provides best results only when the modality filter is used.

The importance of the method for detecting salient points for further computation of VW is investigated by Haas et al. in [13]. Three detectors of interest points are compared: points with high response of DoG, a dense sampling using a regular Cartesian grid and centers of mass of ROIs segmented using the superpixel algorithm [24]. For each salient point, SIFT descriptors [22] are used to span the feature space in which the k–means algorithm identifies clusters and their centers as VWs. Inter–image distances computed as the χ^2 distance between the histograms of the occurrences of the various VWs over all salient points. The methods are evaluated on two datasets. The first evaluation uses the ImageCLEF 2009 medical image annotation data set and error evaluation[4] [32]. Error scores show the improved classification performance achieved by superpixel interest points when compared to dense sampling and DoG detectors. The methods are also evaluated on their ability to locate slices of lung CTs based on the position of the retrieved images. The position is determined by the median position value of the ten best retrieved images per query. Again, the superpixel approach outperforms dense sampling and DoG in terms of sum of squared errors of the vertical positions.

Cross–modality retrieval of ROIs consisting of bounding boxes of organs is addressed by Venkatraghavan et al. in [16]. First, a coarse localization of the regions in whole–body CT and magnetic resonance imaging (MRI) scans is obtained using sped–up robust features (SURF) [33] descriptors in Gabor–filtered images to determine the anatomical region (i.e., cranial, thoracic, abdominal, sacro–lumbar and the extremities). This initial localization is then refined using a fuzzy uniformity index from texture descriptors obtained with 3D local binary patterns computed from a Gabor–filtered 3D volume. A sliding window search is used to exhaustively index volumetric datasets. Inverted files are used to index organs based on a semantic vocabulary containing 14 terms corresponding to organs (e.g., liver, left lung, right lung). The proposed methods are evaluated on a dataset of CT and MRI images from various body regions in terms of localization errors of the bounding boxes of the organs. A comparison with regression and decision forests shows that the proposed methods allow for a more precise localization of the organs. No quantitative evaluation of the retrieval performance is carried out.

---

[4] http://www.idiap.ch/clef2009/evaluation_tools/error_evaluation.pdf

Speed efficiency of CBIR is investigated by Donner et al. in [17] by using small versions of 2D and 3D images. 2D images from the ImageCLEF 2009 medical annotation data set [32] are downsampled to 32×32 thumbnails. 3D CT scans are downsampled to $16 \times 16 \times 16$ volumes. Various retrieval approaches are compared. First, PCA is applied to obtain a more compact representation, in which k–nearest–neighbors use a Euclidean distance and kD–trees to speed up the retrieval process. Second, correlations of rigidly aligned thumbnails are used as a distance metric between the images, which is only carried out in the axial plane for 3D volumes. Third, feature vectors obtained from distribution fields [34] are compared using an l_1–norm distance measure. At last, a histogram of oriented gradients (HOG) from SIFT descriptors extracted with either dense sampling or salient points detected with DoGs are used together with a χ^2 distance. The latter proved to outperform the others using the 2D ImageCLEF 2009 dataset. For 3D CT scans, the retrieval performance is evaluated as the distance between the center of the query image and the center of the most similar image. Again, the approach based on the histograms of gradients is providing best performance. No quantitative evaluation of the retrieval speed is carried out but the authors show that good retrieval performance can be achieved using only highly downsampled image thumbnails.

Clustering efficiency for the computation of VWs is addressed by Pauly et al. in [14] by using multiple random partitioning of the feature space based on extreme random subspace projection ferns. Local appearance with local distributions of color/intensity gradient directions are encoded using color/intensity values, LBP and HOG in 17×17 patches extracted at random locations. In this feature space, random ferns [35] perform a hierarchical partitioning with only one decision function per level. This allows to efficiently categorize an instance with a hierarchical set of binary tests. The authors modify the initial algorithms by using random splits at each level. Bags of VWs are obtained by a concatenation of the VWs from all partitions. The approach is evaluated on the modality classification task at ImageCLEF 2010 [30] and shows an improved accuracy and efficiency when compared to the k–means clustering algorithm.

Purely text–based image retrieval is proposed by Mata et al. in [36], where image retrieval tasks from the ImageCLEF 2009 and 2010 collections are used. The text queries are expanded using the MeSH controlled vocabulary of the National Library of Medicine. The queries are initially divided into N–grams. Then, various query expansion strategies are investigated. The first strategy uses cross–referencing functions provided by the MeSH vocabulary, such as *SeeRelatedDescriptor* and *ConsiderAlso*. The former associates the descriptor with other descriptors and the second returns terms having related linguistic roots. The second strategy uses directly the children terms provided by the hierarchical structure of the MeSH tree. The third strategy uses synonyms (so–called "entry terms" by MeSH) to expand the queries. The Lucene[5] text search engine is used to index and retrieve image documents. The results show that whereas the strategy based on the children terms for expansion performs best on the

---

[5] http://lucene.apache.org/java/docs/index.html

2009 dataset, no significant improvement is achieved in terms of mean average precision when compared with the baseline consisting of using the initial query without expansion.

3 Conclusions

The second workshop on content–based medical retrieval for clinical decision support showed a set of state-of-the-art works, from academic as well as industrial perspectives. We have found that the audience increased from previous meetings at the same venue. The domain is gaining interest in the MICCAI community.

One evident trend is the desire to show the contribution of the field in clinical practice. As such, there is a desire to view and integrate content–based retrieval as one part of larger information access and management systems. Hospital information systems (HIS) and Picture Archival and Communication Systems (PACS) are the backbone of the hospital enterprise. The stored data are the asset for the hospitals and being able to use the knowledge stored in these archives is a key component to case–based reasoning and medical decision–making processes. Image analysis is a single component of this process. Text mining and general textual information retrieval are other components that need to be combined with visual retrieval for decision support in clinical settings. After the lively discussion at this workshop, we plan to continue holding the workshop at future MICCAI conferences, and we hope to see you all there.

Acknowledgments. We would like to thank the EU FP7 projects Khresmoi (257528), Promise (258191) and Chorus+ (249008) for their support as well as the Swiss national science foundation with the MANY project (number 205321–130046). A special thank you also belongs to IBM, which supported the workshop. We would also like to thank all reviewers as they helped assuring the high quality of the presented papers.

References

1. Müller, H., Michoux, N., Bandon, D., Geissbuhler, A.: A review of content–based image retrieval systems in medicine–clinical benefits and future directions. International Journal of Medical Informatics 73(1), 1–23 (2004)
2. Tagare, H.D., Jaffe, C., Duncan, J.: Medical image databases: A content–based retrieval approach. Journal of the American Medical Informatics Association 4(3), 184–198 (1997)
3. Hersh, W., Müller, H., Kalpathy-Cramer, J., Kim, E., Zhou, X.: The consolidated ImageCLEFmed medical image retrieval task test collection. Journal of Digital Imaging 22(6), 648–655 (2009)
4. Aisen, A.M., Broderick, L.S., Winer-Muram, H., Brodley, C.E., Kak, A.C., Pavlopoulou, C., Dy, J., Shyu, C.R., Marchiori, A.: Automated storage and retrieval of thin–section CT images to assist diagnosis: System description and preliminary assessment. Radiology 228(1), 265–270 (2003)

5. Müller, H., Kalpathy–Cramer, J., Caputo, B., Syeda-Mahmood, T., Wang, F.: Overview of the First Workshop on Medical Content–Based Retrieval for Clinical Decision Support at MICCAI 2009. In: Caputo, B., Müller, H., Syeda-Mahmood, T., Duncan, J.S., Wang, F., Kalpathy-Cramer, J. (eds.) MCBR-CDS 2009. LNCS, vol. 5853, pp. 1–17. Springer, Heidelberg (2010)

6. Müller, H., Clough, P., Deselaers, T., Caputo, B. (eds.): ImageCLEF – Experimental Evaluation in Visual Information Retrieval. The Springer International Series On Information Retrieval, vol. 32. Springer, Heidelberg (2010)

7. Müller, H., Deselaers, T., Kim, E., Kalpathy-Cramer, J., Deserno, T.M., Clough, P., Hersh, W.: Overview of the ImageCLEFmed 2007 medical retrieval and annotation tasks. In: Working Notes of the 2007 CLEF Workshop, Budapest, Hungary (September 2007)

8. Kalpathy-Cramer, J., Müller, H., Bedrick, S., Eggel, I., de Herrera, A.S., Tsikrika, T.: The CLEF 2011 medical image retrieval and classification tasks. In: Working Notes of CLEF 2011 (Cross Language Evaluation Forum) (September 2011)

9. André, B., Vercauteren, T., Ayache, N.: Content-Based Retrieval in Endomicroscopy: Toward an Efficient Smart Atlas for Clinical Diagnosis. In: Müller, H., Greenspan, H., Syeda-Mahmood, T. (eds.) MCBR-CDS 2011. LNCS, vol. 7075, pp. 12–23. Springer, Heidelberg (2011)

10. Burner, A., Donner, R., Mayerhoefer, M., Holzer, M., Kainberger, F., Langs, G.: Texture Bags: Anomaly Retrieval in Medical Images Based on Local 3D-Texture Similarity. In: Müller, H., Greenspan, H., Syeda-Mahmood, T. (eds.) MCBR-CDS 2011. LNCS, vol. 7075, pp. 116–127. Springer, Heidelberg (2011)

11. Foncubierta-Rodríguez, A., Depeursinge, A., Müller, H.: Using Multiscale Visual Words for Lung Texture Classification and Retrieval. In: Greenspan, H., Müller, H., Syeda Mahmood, T. (eds.) MCBR-CDS 2011. LNCS, vol. 7075, pp. 69–79. Springer, Heidelberg (2011)

12. Vanegas, J.A., Caicedo, J.C., González, F.A., Romero, E.: Histology Image Indexing using a Non–Negative Semantic Embedding. In: Greenspan, H., Müller, H., Syeda-Mahmood, T. (eds.) MCBR-CDS 2011. LNCS, vol. 7075, pp. 80–91. Springer, Heidelberg (2011)

13. Haas, S., Donner, R., Burner, A., Holzer, M., Langs, G.: Superpixel-Based Interest Points for Effective Bags of Visual Words Medical Image Retrieval. In: Greenspan, H., Müller, H., Syeda-Mahmood, T. (eds.) MCBR-CDS 2011. LNCS, vol. 7075, pp. 58–68. Springer, Heidelberg (2011)

14. Pauly, O., Mateus, D., Navab, N.: Building Implicit Dictionaries Based on Extreme Random Clustering for Modality Recognition. In: Greenspan, H., Müller, H., Syeda-Mahmood, T. (eds.) MCBR-CDS 2011. LNCS, vol. 7075, pp. 47–57. Springer, Heidelberg (2011)

15. Costa, M.J., Tsymbal, A., Hammon, M., Cavallaro, A., Sühling, M., Seifert, S., Comaniciu, D.: A Discriminative Distance Learning–Based CBIR Framework for Characterization of Indeterminate Liver Lesions. In: Greenspan, H., Müller, H., Syeda-Mahmood, T. (eds.) MCBR-CDS 2011. LNCS, vol. 7075, pp. 92–104. Springer, Heidelberg (2011)

16. Venkatraghavan, V., Ranjan, S.: Semantic Analysis of 3D Anatomical Medical Images for Sub–Image Retrieval. In: Greenspan, H., Müller, H., Syeda-Mahmood, T. (eds.) MCBR-CDS 2011. LNCS, vol. 7075, pp. 139–151. Springer, Heidelberg (2011)

17. Donner, R., Haas, S., Burner, A., Holzer, M., Bischof, H., Langs, G.: Evaluation of fast 2D and 3D Medical Image Retrieval Approaches Based on Image Miniatures. In: Greenspan, H., Müller, H., Syeda-Mahmood, T. (eds.) MCBR-CDS 2011. LNCS, vol. 7075, pp. 128–138. Springer, Heidelberg (2011)
18. Avni, U., Greenspan, H., Konen, E., Sharon, M., Goldberger, J.: X–ray categorization and retrieval on the organ and pathology level, using patch–based visual words. IEEE Transactions on Medical Imaging 30(3), 733–746 (2011)
19. Rahman, M.M., Antani, S.K., Fushman, D.D., Thoma, G.R.: Biomedical Image Retrieval using Multimodal Context and Concept Feature Spaces. In: Greenspan, H., Müller, H., Syeda-Mahmood, T. (eds.) MCBR-CDS 2011. LNCS, vol. 7075, pp. 24–35. Springer, Heidelberg (2011)
20. Vercauteren, T., Perchant, A., Malandain, G., Pennec, X., Ayache, N.: Robust mosaicing with correction of motion distortions and tissue deformations for in vivo fibered microscopy. Medical Image Analysis 10(5), 673–692 (2006)
21. Sivic, J., Zisserman, A.: Video Google: Efficient Visual Search of Videos. In: Ponce, J., Hebert, M., Schmid, C., Zisserman, A. (eds.) Toward Category-Level Object Recognition. LNCS, vol. 4170, pp. 127–144. Springer, Heidelberg (2006)
22. Lowe, D.G.: Distinctive image features from scale-invariant keypoints. International Journal of Computer Vision 60(2), 91–110 (2004)
23. Ojala, T., Pietikainen, M., Maenpaa, T.: Multiresolution gray–scale and rotation invariant texture classification with local binary patterns. IEEE Transactions on Pattern Analysis and Machine Intelligence 24(7), 971–987 (2002)
24. Wildenauer, H., Mičušík, B., Vincze, M.: Efficient Texture Representation Using Multi-Scale Regions. In: Yagi, Y., Kang, S.B., Kweon, I.S., Zha, H. (eds.) ACCV 2007, Part I. LNCS, vol. 4843, pp. 65–74. Springer, Heidelberg (2007)
25. Ling, H., Okada, K.: Diffusion distance for histogram comparison. In: 2006 IEEE Computer Society Conference on Computer Vision and Pattern Recognition, vol. 1, pp. 246–253 (June 2006)
26. Breiman, L.: Random forests. Machine Learning 45(1), 5–32 (2001)
27. Safi, A., Baust, M., Pauly, O., Castaneda, V., Lasser, T., Mateus, D., Navab, N., Hein, R., Ziai, M.: Computer–Aided Diagnosis of Pigmented Skin Dermoscopic Images. In: Greenspan, H., Müller, H., Syeda-Mahmood, T. (eds.) MCBR-CDS 2011. LNCS, vol. 7075, pp. 105–115. Springer, Heidelberg (2011)
28. Li, F., Shen, C., Li, C.: Multiphase soft segmentation with total variation and H^1 regularization. Journal of Mathematical Imaging and Vision 37(2), 98–111 (2010)
29. Cichocki, A., Zdunek, R., Amari, S.: New algorithms for non–negative matrix factorization in applications to blind source separation. In: 2006 IEEE International Conference on Acoustics, Speech and Signal Processing, ICASSP 2006 Proceedings, vol. 5, pp. 621–624 (May 2006)
30. Müller, H., Kalpathy-Cramer, J., Eggel, I., Bedrick, S., Said, R., Bakke, B., Kahn Jr, C.E., Hersh, W.: Overview of the CLEF 2010 medical image retrieval track. In: Working Notes of CLEF 2010 (Cross Language Evaluation Forum) (September 2010)
31. Baeza Yates, R.A., Neto, B.R.: Modern Information Retrieval. Addison-Wesley Longman Publishing Co., Inc., Boston (1999)
32. Tommasi, T., Caputo, B., Welter, P., Güld, M.O., Deserno, T.M.: Overview of the CLEF 2009 Medical Image Annotation Track. In: Peters, C., Caputo, B., Gonzalo, J., Jones, G.J.F., Kalpathy-Cramer, J., Müller, H., Tsikrika, T. (eds.) CLEF 2009. LNCS, vol. 6242, pp. 85–93. Springer, Heidelberg (2010)

33. Bay, H., Ess, A., Tuytelaars, T., Gool, L.V.: Speeded–up robust features (surf). Computer Vision and Image Understanding 110(3), 346–359 (2008)
34. Sevilla, L., Learned-Miller, E.: Distribution fields. Technical Report UM–CS–2011–027, Dept. of Computer Science, University of Massachusetts Amherst (2011)
35. Özuysal, M., Calonder, M., Lepetit, V., Fua, P.: Fast keypoint recognition using random ferns. IEEE Transactions on Pattern Analysis and Machine Intelligence 32, 448–461 (2010)
36. Mata, J., Crespo, M., Maña, M.J.: Using MeSH to Expand Queries in Medical Image Retrieval. In: Greenspan, H., Müller, H., Syeda-Mahmood, T. (eds.) MCBR-CDS 2011. LNCS, vol. 7075, pp. 36–46. Springer, Heidelberg (2011)

Content-Based Retrieval in Endomicroscopy: Toward an Efficient *Smart Atlas* for Clinical Diagnosis

Barbara André[1,2], Tom Vercauteren[1], and Nicholas Ayache[2]

[1] Mauna Kea Technologies (MKT), Paris, France
[2] INRIA - Asclepios Research Project, Sophia Antipolis

Abstract. In this paper we present the first Content-Based Image Retrieval (CBIR) framework in the field of *in vivo* endomicroscopy, with applications ranging from training support to diagnosis support. We propose to adjust the standard Bag-of-Visual-Words method for the retrieval of endomicroscopic videos. Retrieval performance is evaluated both indirectly from a classification point-of-view, and directly with respect to a perceived similarity ground truth. The proposed method significantly outperforms, on two different endomicroscopy databases, several state-of-the-art methods in CBIR. With the aim of building a self-training simulator, we use retrieval results to estimate the interpretation *difficulty* experienced by the endoscopists. Finally, by incorporating clinical knowledge about perceived similarity and endomicroscopy semantics, we are able: 1) to learn an adequate visual similarity distance and 2) to build visual-word-based *semantic* signatures that extract, from low-level visual features, a higher-level clinical knowledge expressed in the endoscopist own language.

1 Introduction

What Is pCLE? Probe-based Confocal Laser Endomicroscopy (pCLE) allows the endoscopists to image the epithelium at a microscopic scale, *in vivo* and *in situ*, at real-time frame rate. Thanks to this novel imaging system illustrated in Fig. 1, the endoscopists have the opportunity to perform non-invasive *optical biopsies*. Traditional biopsies result in histological images that are classically diagnosed *ex vivo* by pathologists and not by endoscopists. The *in vivo* diagnosis of pCLE images is therefore a critical challenge for the endoscopists who typically have less expertise in histopathology. Fig. 2 illustrates this challenge by showing the high variability in appearance of pCLE mosaics of colonic polyps. Therefore, our main goal is to assist the endoscopists in the *in vivo* interpretation of pCLE image sequences.

Why Using CBIR to Support pCLE Diagnosis? When establishing a diagnosis, physicians typically rely on similarity-based reasoning. To mimic this

H. Müller et al. (Eds.): MCBR-CDS 2011, LNCS 7075, pp. 12–23, 2012.

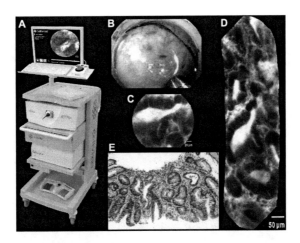

Fig. 1. **(A)** Setup of pCLE imaging system. **(B)** Endoscopic image of a colonic polyp diagnosed with tubular adenoma (macroscopic view), and the pCLE miniprobe. **(C)** Acquired pCLE image of the colonic polyp (microscopic "en-face" view, i.e. frontal view). **(D)** pCLE mosaic image associated to the acquired pCLE video. **(E)** Histopathology image of the colonic polyp (microscopic transverse view), obtained from a traditional biopsy corresponding to the "optical biopsy" site.

process, we explore content-based image retrieval (CBIR) approaches for diagnosis support. Our main objective is to develop a system which automatically extracts, from a training database, several videos that are visually similar to the pCLE video of interest, but that are annotated with metadata such as textual diagnosis. Such a retrieval system, acting like a *Smart Atlas* that opens a comprehensive book at the right pages, should help the endoscopist in making an informed decision and therefore a more accurate pCLE diagnosis. Another relevant application of our retrieval framework, which is presented in this paper, is to build a self-training simulator that features difficulty selection, in order to help the endoscopists in shortening their learning curve in pCLE diagnosis.

2   Adjusting Bag of Visual Words for pCLE Retrieval

Toward a Dense Bag-of-Visual-Words Method. The Bag-of-Visual-Words method, proposed by Sivic and Zisserman [1], is a CBIR method which has been successfully applied in computer vision, in particular by Zhang et al. [2] for the classification of texture images. Another CBIR method based on multi-scale affine kernels is proposed by Sedan-Mahmud et al. in [3], with the purpose of object categorization. Noticing that pCLE images have a similar appearance to texture images and that their discriminative information is densely distributed, we propose in [4] a dense Bag-of-Visual-Words method for pCLE retrieval. Our dense Bag-of-Visual-Words method consists of: **1)** the detection of disk regions in the images at each point of a dense regular grid, **2)** the description of these

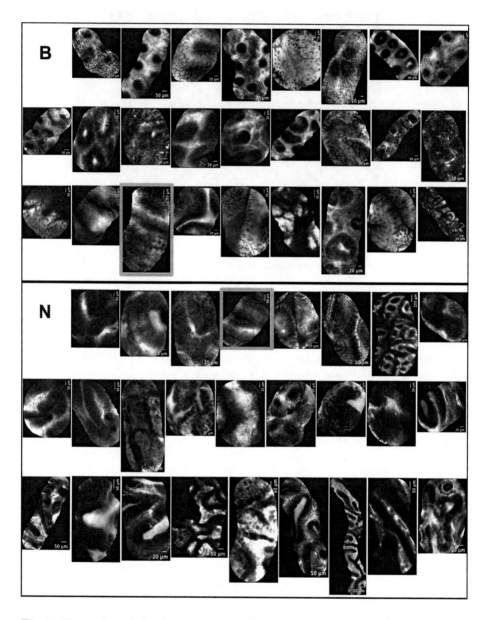

Fig. 2. Illustration of the *Semantic Gap* between low-level visual features and high-level clinical knowledge, on pCLE mosaics of the colonic polyps database. Scale bars provide a cue on the field of view size. On top (resp. bottom) are the mosaics of the polyps diagnosed as non-neoplastic (resp. neoplastic) indicated by **B** (resp. **N**). The closer to the boundary the mosaics are, the less obvious is their diagnosis according to their visual appearance. The two framed mosaics might look similar although they belong to different pathological classes.

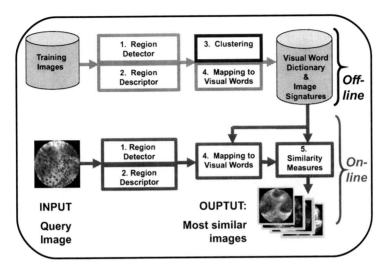

Fig. 3. Overview of the retrieval pipeline

regions into description vectors using the Scale Invariant Feature Transform (SIFT), **3)** the K-means clustering of all description vectors into K visual words, **4)** the construction for each image of a visual word histogram, called *visual* signature, that is invariant with respect to viewpoint changes (translations and rotations) and illumination changes (affine transformations of the intensity). The whole retrieval pipeline is illustrated in Fig. 3.

Video Retrieval Using Explicit or Implicit Mosaics. As the pCLE miniprobe moves in constant contact with the tissue, the successive images from the acquired pCLE video are mostly related by viewpoint changes. We thus use the video mosaicing technique of Vercauteren et al. [5] that employs non-rigid registration to project the temporal dimension of a video sequence onto one mosaic image with a larger field of view and of higher resolution. Some examples of the resulting mosaic images are presented in Fig. 2. In [4], we propose two different representations of pCLE videos depending on the time constraints. In the *explicit mosaic* representation, mosaic images are built with the time-consuming non-rigid registration, then we compute the *visual* signature for each mosaic image, finally the video signature is obtained using a histogram summation technique. In the *implicit mosaic* representation, we first compute the *visual* signature for each single image, then we leverage coarse registration results between time-related images, provided by the real-time version of video mosaicing, in order to apply overlap weighting to the visual words before performing the histogram summation step. We define the similarity distance between two pCLE videos as the χ^2 pseudo-distance between their signatures. The resulting pCLE video retrieval methods were applied on a database of colonic polyps, but also on a database of Barrett's esophagus as illustrated in Fig. 4.

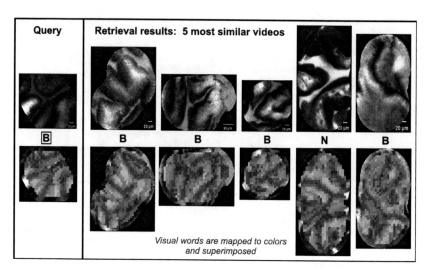

Fig. 4. Typical retrieval results on the Barrett's esophagus. The video query on the left is followed by its $k = 5$ nearest neighbors on the right. The pCLE video sequences are represented by their corresponding fused mosaic image. The superimposed visual words, mapped to colors, are highlighting the geometrical structures observed in the pCLE mosaic images. **B** indicates *benign* (i.e. non-neoplastic) and **N** *neoplastic*.

3   Evaluating pCLE Retrieval Performance

Indirect Retrieval Evaluation Based on Classification. Due to the inherent difficulty to obtain a ground truth for CBIR on biomedical applications, as pointed out by Müller et al. [6] and by Akgül et al. [7] , it may be advantageous to evaluate retrieval performance in an indirect manner using classification. We consider several pathological classes, for example two classes, and perform k-nearest neighbor classification with leave-one-patient-out (LOPO) cross-validation. This allow us to compare, on the different pCLE databases, the classification performances of our "Dense-Sift" method with respect to several state-of-the-art CBIR methods, namely the statistics-based "Haralick" method [8], the dense "Textons" method [9], the sparse "HH-Sift" method [2], and the "NBNN" classifier [10]. We refer the reader to [4] for a detailed description of these four methods and the evaluation methodology. We demonstrate that, in terms of classification performance, our "Dense-Sift" method significantly outperforms the state-of-the-art methods, on the pCLE database of colonic polyps, as illustrated in Fig. 5, but also on the pCLE database of Barrett's esophagus.

Building a Ground Truth for Perceived Similarity. In order to directly evaluate retrieval methods in terms of visual similarity, we aim at constructing

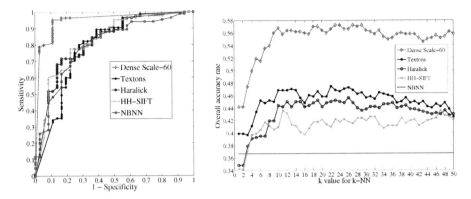

Fig. 5. Left: ROC curves at $k = 5$ neighbors from the LOPO binary classification of pCLE videos of colonic polyps. **Right:** LOPO 5-class classification of pCLE mosaics by the methods on the colonic polyps database. The NBNN does not depend on k.

a ground truth for the perceived similarity between pCLE videos, which is a difficult task because of the subjective appreciation of visual similarities. To facilitate the ground truth generation, we develop in [11] an online survey tool presented in Fig. 6 and available at `http://smartatlas.maunakeatech.com`, login: MICCAI-User, password: MICCAI2011. This "Visual Similarity Scoring" (VSS) tool allows multiple endoscopists to individually evaluate the visual similarity that they perceived between pCLE videos, according to the four-point Likert scale: "very dissimilar" (L=-2), "rather dissimilar" (L=-1), "rather similar" (L=+1), "very similar" (L=+2). 17 observers, ranging from middle expert to expert in pCLE diagnosis, performed as many scoring processes as they could. In total, $4,836$ similarity scores were given for $2,178$ distinct video couples. 16.2% of all $13,434$ distinct video couples were scored, thus composing a sparse but representative ground truth for perceived similarity.

Direct Retrieval Evaluation against Perceived Similarity. From the sparse perceived similarity ground truth obtained on the database of colonic polyps, we are able to perform direct retrieval evaluation by measuring the correlation between the similarity distance based on *visual* signatures and the true perceived similarity. We demonstrate in [11] that, in terms of correlation with the perceived similarity, our "Dense-Sift" method also significantly outperforms the state-of-the-art methods. Furthermore, we define *sparse recall* curves by considering the percentage of videos perceived as "very similar" to the query that were found in the k-neighborhood of the query by the retrieval methods. In Fig. 7 on the left, the *sparse recall* curve of "Dense-Sift" is significantly above the *sparse recall* curves of the state-of-the-art methods, which confirms the previous indirect comparison results.

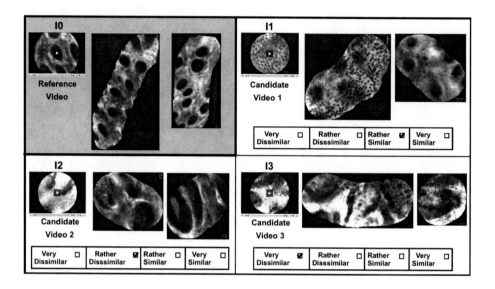

Fig. 6. Schematic outline of the online "Visual Similarity Scoring" tool showing the example of a scoring process, where 3 video couples (I_0, I_1), (I_0, I_2) and (I_0, I_3) are proposed. Each video of the colonic polyps database is summarized by a set of mosaic images, even though the video remains available for viewing.

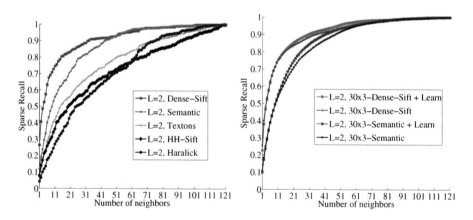

Fig. 7. *Sparse recall* curves on the colonic polyps database, showing the ability of the retrieval methods to capture video pairs perceived as "very similar". **Left:** without cross-validation, without distance learning. **Right:** Median of *sparse recall* curves obtained with 30×3-fold cross-validation, before and after distance learning.

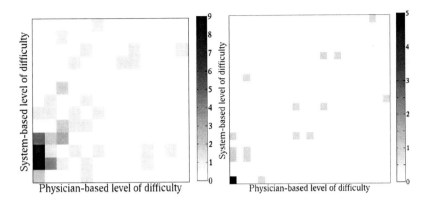

Fig. 8. Joint histograms for the colonic polyps database **(left)**, and for the Barrett's esophagus database **(right)**; x-axis is the *difficulty* experienced by all the physicians and y-axis is our estimated *difficulty*. On the colonic polyps database, 11 physicians, 3 expert and 8 non expert, individually diagnosed 63 videos. On the Barrett's esophagus database, 21 physicians, 9 expert and 12 non expert, individually diagnosed 20 videos.

4   Estimating the Interpretation Difficulty

With the aim of building a self-training simulator for pCLE diagnosis with an adjustable level of difficulty, we propose in [12] to automatically estimate the interpretation *difficulty* associated to a pCLE video by exploiting our retrieval results. For the *difficulty* estimation, we include two main *difficulty* attributes: the first attribute reflects the contextual discrepancies between the video query and its similarity neighborhood, and the second attribute measures the intrinsic ambiguity of the video query with respect to the two pathological classes. Using a robust linear regression model, we leverage the experienced *difficulty* to learn the difficulty predictor from these two attributes. For the learning and the validation steps, another type of ground truth is needed. This ground truth is the *difficulty* experienced by the endoscopists, which is given by the percentage of false pCLE diagnoses, with respect to histology, among several endoscopists. We demonstrate, using permutation tests, that the correlation between the estimated *difficulty* and the experienced *difficulty* is statistically significant on both pCLE databases. Joint histograms can be qualitatively appreciated in Fig. 8. We also notice that correlation coefficients are higher when considering only the non-expert endoscopists.

5   Learning pCLE Semantic and Similarity Distance

Learning pCLE Semantics. Aiming at providing the endoscopists with semantic insight into the retrieval results, we investigate pCLE semantic learning. Our semantic ground truth contains 8 binary semantic concepts that are illustrated in Fig. 9 on the left. These semantic concepts were defined by experts in

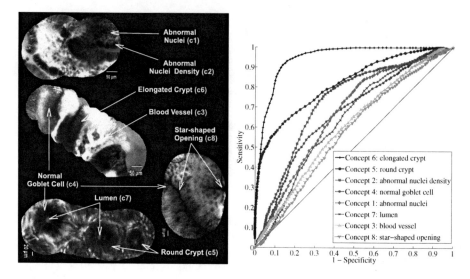

Fig. 9. Left: Examples of training pCLE videos of the colonic polyps database, represented by mosaic images and annotated with the 8 semantic concepts. The two mosaics on the top mosaics show neoplastic colonic polyp, while the two mosaics on the bottom show non-neoplastic colonic polyps. **Right:** ROC curves showing, for each semantic concept, the classification performance of the visual-word-based *semantic* signature. Each ROC curve associated to a concept c_j is computed with cross-validation by thresholding on the *semantic weight* s_j.

pCLE in order to support the *in vivo* diagnosis of colonic polyps [13]. Rasiwasia et al. [14] propose a probabilistic method which estimates for each semantic concept the probability that, given a visual feature vector in an image, the semantic concept is present in the image. In [15], Kwitt et al. apply this method for learning pit pattern concepts in endoscopic images of colonic polyps. These pit pattern concepts at the macroscopic level can be seen as corresponding to our semantic concepts at the microscopic level. In [16], we present a Fisher-based approach that leverages the ground-truth data about pCLE semantics in order to compute visual-word-based *semantic* signatures. For each video, our learned *semantic* signature contains 8 floating values, called *semantic weight* $s_j^j \in \{1,..,8\}$, such that each *semantic weight* s_j reflects how much the presence of the semantic concept c_j is expressed by the visual words describing the video. Fig. 9 on the right shows the classification performance of the *semantic* signatures with a ROC curve for each semantic concept. The fact that the ROC curves are above the diagonal indicates that each semantic concept contributes to the relevance of the *semantic* signature. Contrary to the "Dense-Sift" *visual* signatures, the *semantic* signatures are not employed for the retrieval task but they are used as an additional information which provides a semantic translation of the visual retrieval outputs. We generate a *sparse recall* curve for the *semantic* signatures

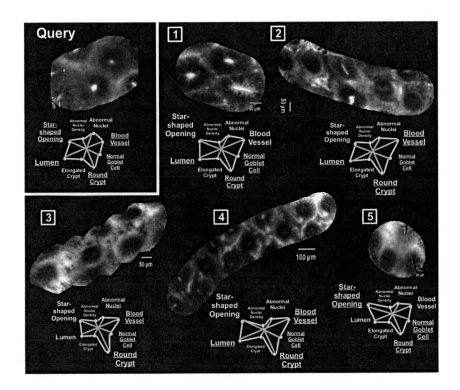

Fig. 10. Typical pCLE retrieval results with semantic extraction on the colonic polyps database, from a non-neoplastic query. The font size of each written semantic concept is proportional to the automatically computed value of the concept coordinate in the star plot. Underlined concepts are those which were annotated as present in the semantic ground truth. (In practice, the semantic ground truth is not known for the video query, but it is disclosed here for illustration purposes.)

as a means to evaluate them from the perceptual point of view and to check their consistency with respect to the *visual* signature. This consistency is shown in Fig. 7 by the fact that the *sparse recall* curve of the *semantic* signature is much closer to the curve of the "Dense-Sift" *visual* signature than the curves of state-of-the-art methods. In Fig. 10, we present a typical retrieval result, where the most similar pCLE videos have been extracted using the *visual* signatures of "Dense-Sift", and where the additional semantic information is provided using a star-plot representation of each visual-word-based *semantic* signature.

Learning Similarity Distance between pCLE Videos. In order to boost retrieval performance, we propose in [11,16] a method to learn an adjusted similarity distance from the perceived similarity ground truth. In a similar way to the method of Philbin et al. [17], our strategy is to shorten the distances between "very similar" videos and to increase the distances between "non very similar" videos. A linear transformation of video signatures is optimized, that minimizes a margin-based cost function differentiating "very similar" video pairs from the others. The *sparse recall* curve associated the transformed *visual* signature is shown in Fig. 7 on the right. We demonstrate that, in terms of correlation with the perceived similarity, the distance learning method allows to improve with statistical significance, the correlation with the perceived similarity. For consistency checking, we also test the distance learning method on the *semantic* signatures, and we show that the transformed *semantics* signature provides a significantly higher correlation with the perceived similarity.

6   Conclusion

In this paper, we presented the first CBIR framework for training and diagnosis support in the field of *in vivo* endomicroscopy. Our main contributions are: **1)** a dense Bag-of-Visual-Words method for a pCLE video retrieval that outperforms several state-of-the-art methods, **2)** the construction of a perceived similarity ground truth, **3)** the development of objective and generic tools for retrieval evaluation, **4)** the estimation of pCLE interpretation *difficulty*, **5)** a method for pCLE semantics learning to provide the endoscopists with a semantic insight into the retrieval results, **6)** a method for perceived similarity learning to boost pCLE retrieval. The resulting pCLE retrieval system, augmented with the estimation of non-visual features such as interpretation difficulty and semantic concepts, is our proposed "Smart Atlas". The clinical applications of the "Smart Atlas" include pCLE diagnosis support, training support and knowledge sharing.

Future work will focus on the clinical validation of the "Smart Atlas" and its application to other organs and pathologies. We also plan to investigate spatio-temporal retrieval and to enrich the databases, for example with other metadata to allow for multimodal information retrieval. The "Smart Atlas" tool should allow to increase the diagnostic performance of the endoscopists and to reduce interobserver variability, which should ultimately improve patient outcomes.

Acknowlegments. We would like to thank Pr. Michael B. Wallace and Dr. Anna M. Buchner, who have acquired, analyzed and annotated all the pCLE videos, at the Mayo Clinic in Jacksonville (Florida, US), for their precious contributions.

References

1. Sivic, J., Zisserman, A.: Efficient visual search of videos cast as text retrieval. IEEE Transactions on Pattern Analysis and Machine Intelligence 31(4), 591–606 (2009)
2. Zhang, J., Lazebnik, S., Schmid, C.: Local features and kernels for classification of texture and object categories: a comprehensive study. International Journal of Computer Vision 73, 213–238 (2007)
3. Syeda-Mahmood, T.F., Wang, F., Beymer, D.: Recognition of object categories using affine kernels. In: Multimedia Information Retrieval, pp. 15–24 (2010)

4. André, B., Vercauteren, T., Buchner, A.M., Wallace, M.B., Ayache, N.: A smart atlas for endomicroscopy using automated video retrieval. Medical Image Analysis 15(4), 460–476 (2011)
5. Vercauteren, T., Perchant, A., Malandain, G., Pennec, X., Ayache, N.: Robust mosaicing with correction of motion distortions and tissue deformation for in vivo fibered microscopy. Medical Image Analysis 10(5), 673–692 (2006)
6. Müller, H., Kalpathy-Cramer, J., Eggel, I., Bedrick, S., Reisetter, J., Kahn, C.E., Hersh, W.R.: Overview of the clef 2010 medical image retrieval track. In: CLEF (Notebook Papers/LABs/Workshops) (2010)
7. Akgül, C.B., Rubin, D.L., Napel, S., Beaulieu, C.F., Greenspan, H., Acar, B.: Content-based image retrieval in radiology: Current status and future directions. Journal of Digital Imaging 24(2), 208–222 (2011)
8. Haralick, R.M.: Statistical and structural approaches to texture. Proceedings of the IEEE 67, 786–804 (1979)
9. Leung, T., Malik, J.: Representing and recognizing the visual appearance of materials using three-dimensional textons. International Journal of Computer Vision 43, 29–44 (2001)
10. Boiman, O., Shechtman, E., Irani, M.: In defense of nearest-neighbor based image classification. In: Proceedings of the IEEE Conference on Computer Vision and Pattern Recognition (CVPR 2008), pp. 1–8 (2008)
11. André, B., Vercauteren, T., Buchner, A.M., Wallace, M.B., Ayache, N.: Retrieval Evaluation and Distance Learning from Perceived Similarity between Endomicroscopy Videos. In: Fichtinger, G., Martel, A., Peters, T. (eds.) MICCAI 2011, Part III. LNCS, vol. 6893, pp. 297–304. Springer, Heidelberg (2011)
12. André, B., Vercauteren, T., Buchner, A.M., Shahid, M.W., Wallace, M.B., Ayache, N.: An Image Retrieval Approach to Setup Difficulty Levels in Training Systems for Endomicroscopy Diagnosis. In: Jiang, T., Navab, N., Pluim, J.P.W., Viergever, M.A. (eds.) MICCAI 2010. LNCS, vol. 6362, pp. 480–487. Springer, Heidelberg (2010)
13. Kiesslich, R., Burg, J., Vieth, M., Gnaendiger, J., Enders, M., Delaney, P., Polglase, A., McLaren, W., Janell, D., Thomas, S., Nafe, B., Galle, P.R., Neurath, M.F.: Confocal laser endoscopy for diagnosing intraepithelial neoplasias and colorectal cancer in vivo. Gastroenterology 127(3), 706–713 (2004)
14. Rasiwasia, N., Moreno, P.J., Vasconcelos, N.: Bridging the gap: Query by semantic example. IEEE Transactions on Multimedia 9(5), 923–938 (2007)
15. Kwitt, R., Rasiwasia, N., Vasconcelos, N., Uhl, A., Häfner, M., Wrba, F.: Learning Pit Pattern Concepts for Gastroenterological Training. In: Fichtinger, G., Martel, A., Peters, T. (eds.) MICCAI 2011, Part III. LNCS, vol. 6893, pp. 280–287. Springer, Heidelberg (2011)
16. André, B., Vercauteren, T., Buchner, A.M., Wallace, M.B., Ayache, N.: Learning semantic and visual similarity for endomicroscopy video retrieval. INRIA Technical Report RR-7722, INRIA (August 2011)
17. Philbin, J., Isard, M., Sivic, J., Zisserman, A.: Descriptor learning for efficient retrieval. In: Daniilidis, K. (ed.) ECCV 2010, Part III. LNCS, vol. 6313, pp. 677–691. Springer, Heidelberg (2010)

Biomedical Image Retrieval Using Multimodal Context and Concept Feature Spaces

Md. Mahmudur Rahman, Sameer K. Antani, Dina Demner Fushman, and George R. Thoma

U.S. National Library of Medicine,
National Institutes of Health, Bethesda, MD, USA
{rahmanmm,santani,ddemner,gthoma}@mail.nih.gov

Abstract. This paper presents a unified medical image retrieval method that integrates visual features and text keywords using multimodal classification and filtering. For content-based image search, concepts derived from visual features are modeled using support vector machine (SVM)-based classification of local patches from local image regions. Text keywords from associated metadata provides the context and are indexed using the vector space model of information retrieval. The concept and context vectors are combined and trained for SVM classification at a global level for image modality (e.g., CT, MR, x-ray, etc.) detection. In this method, the probabilistic outputs from the modality categorization are used to filter images so that the search can be performed only on a candidate subset. An evaluation of the method on ImageCLEFmed 2010 dataset of 77,000 images, XML annotations and topics results in a mean average precision (MAP) score of 0.1125. It demonstrates the effectiveness and efficiency of the proposed multimodal framework compared to using only a single modality or without using any classification information.

1 Introduction

The search for relevant and actionable information is key to achieving clinical and research goals in biomedicine. Biomedical information exists in different forms: as text, illustrations, and images in journal articles, documents, and other collections, and as patient cases in electronic health records. For example, in scientific publications, images are used to elucidate the text and can be easily understood in context. For example, Fig. 1 along with its caption are fairly informative in the context of the paper [1] *"Eosinophilic cellulitis-like reaction to subcutaneous etanercept injection"*. Taken out of context, the caption provides little information about the image, and the image does not provide enough information about the nature of the skin reaction. This example illustrates both the problem of finding text that provides sufficient information about the image without introducing irrelevant information, and the potential benefits of combining information provided by the text and image.

While there is a substantial amount of completed and ongoing research in both the text and content based image retrieval (CBIR) in medical domain, much

H. Müller et al. (Eds.): MCBR-CDS 2011, LNCS 7075, pp. 24–35, 2012.

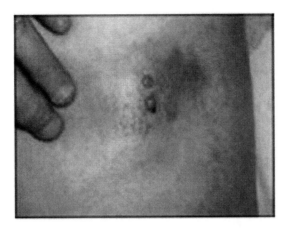

Figure 1: Reaction to intradermal adalimumab 1 to 2 days after the fourth dose

Fig. 1. Example image along with its caption in an article

remains to be done to see how effectively these two approaches can complement each other in an integrated framework. Biomedical image retrieval based on multimodal sources has been only recently gaining popularity due the large amount information sources [2,3]. The results of the past medical retrieval tracks of ImageCLEF[1] suggest that the combination of visual and text based image searches provides better results than using the two different approaches individually.

Previous studies also have shown that imaging modality is an important aspect of medical retrieval [4]. In user-studies, clinicians have indicated that modality is one of the most important search filters that they would like to use. In fact, quality and speed of image retrieval from large biomedical collections can be improved by reducing the search space by filtering out irrelevant images and learning about the image categories. For example, to search "posteroanterior (PA) chest x-rays with enlarged heart", automatically classified images in the collection could be organized according to modality (e.g., x-ray), body part (e.g., chest), and orientation (e.g., PA) criteria. Next, similarity matching can be performed between query and target images in the corresponding filtered subset to find "enlarged heart" as a distinct visual or textual concept. Some medical image search engines, such as Goldminer[2] and Yottalook[3] allow users to limit the search results to a particular modality. However, this modality is typically extracted from the caption and is often not correct or present.

---

[1] http://imageclef.org
[2] http://goldminer.arrs.org/home.php
[3] http://www.yottalook.com

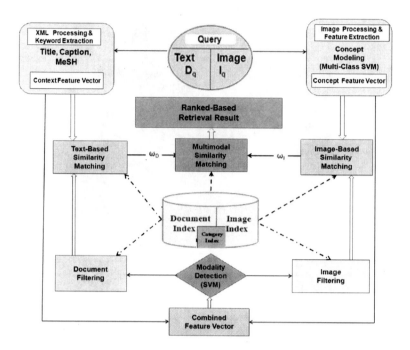

Fig. 2. Process flow diagram of the multimodal retrieval framework

Studies have also shown that the modality can be extracted from the image itself using visual features. For example, in [5], the automatic categorization of 6231 radiological images into 81 categories is examined by utilizing a combination of low-level global texture features with low-resolution scaled images and a K-nearest-neighbors (KNN) classifier. In [6], the performances of two medical image categorization architectures with and without a learning scheme are evaluated on 10,322 images of 33 categories based on modality, body part, and orientation with a high accuracy rate of more than 95%. Although these approaches demonstrated promising results for medical image classification at a global level, they do not relate classification to retrieval in a direct manner, instead only stressed its value for image annotation and pre-filtering purposes.

To minimize limitations of low-level feature representations that result in the semantic gap and motivated by the successful use of machine learning in information retrieval (IR), we present a multimodal classification-based medical image retrieval method. We perform the multimodal search based on image classification and filtering using both textual and visual features. Text feature provide the context while the concept is derived from the visual features. In this framework, the modality specific information that is available as probabilistic outputs of SVM learning on the query and database images is used select the relevany image subset. It is a primary goal of this work to develop improved information retrieval techniques by moving beyond conventional text-based searching to combining both text and visual features extracted from collections of full-text biomedical journal articles, images and illustrations within these, and a collection of patient cases.

Fig. 2 shows the process flow diagram of the proposed multimodal retrieval approach. As can be seen from the top portion of Fig. 2, a search can be initiated simultaneously based on both text (left) and image parts (right) of a multimodal query and later the individual similarity scores are weighted combined (middle) for a final ranked result list. In addition, the text and image features are combined (bottom) to determine the query image modality from a SVM classification sub-system and based on that information only filtered images are accessed from the document and image indexes for further similarity matching.

The proposed approach and an evaluation of its efficacy are presented as follows: in Section 2, we briefly describe the image representation approach in concept and context feature spaces. Section 3 describes the multimodal search approach and Section 4 presents the modality detection and filtering approach based on the SVM classification. The experiments and the analysis of the results are presented in Section 5.

2   Image Feature Representation

The performance of a classification and/or retrieval system depends on the underlying image representation, usually in the form of a feature vector. The following feature vectors are generated at different levels of abstraction.

2.1   Context-Based Image Representation

For purposes of this research we use the ImageCLEFmed 2010 dataset [4] that is provided to the participants of the evaluation. The collection comprises journal articles from two journals published by the Radiological Society of North America (RSNA), viz., *Radiographics* and *Radiology*. The collection includes full text from the articles and all images and figures within these. In all there are nearly 77,500 images from over 5,600 articles. The contents of this collection represent a broad and significant body of medical knowledge, which make the retrieval more challenging. The collection contains a variety of imaging modalities, image sizes, and resolutions and can be considered as a fairly a realistic set for evaluating medical image retrieval techniques.

Each image in the data set is represented as a structured document of image-related text, which is termed as *context* here. Now each image in the collection is attached to a manually annotated case or lab report in a XML file. It is necessary to index these annotation files into an easily accessible representation. There are a variety of indexing techniques which mostly rely on keywords or terms to represent the information content of documents [7]. In our case, information from only relevant tags are extracted and preprocessed by removing stop words that are considered to be of no importance for the actual retrieval process. Subsequently, the remaining words are reduced to their stems, which finally

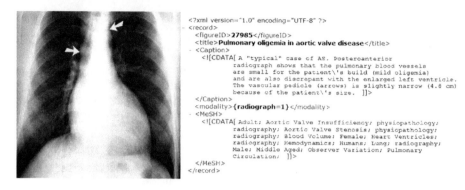

Fig. 3. Sample Chest x-ray image with annotation

form the index terms or keywords of the annotation files. Next, the annotation files (document) are modeled as a vector of words based on the popular vector space model (VSM) of IR [7]. Our representation includes the title, and MeSH terms of the article in which the image appears as well as caption of the images. Fig. 3 shows an example chest x-ray image from the collection along with its annotation which is generated from the article where the image appears.

Let $T = \{t_1, t_2, \cdots, t_N\}$ denote the set of terms in the collection. Then it can represent a document D_j as vector in a N-dimensional space as $\mathbf{f}_j^D = [w_{j1}, w_{j2}, \cdots, w_{jN}]^T$. The element w_{jk} denotes the weight of term t_k in document D_j, depending on its information content. A weighting scheme has two components: a global weight and a local weight. The global importance of a term is indicating its overall importance in the entire collection, weighting all occurrences of the term with the same value. The popular *tf-idf* term-weighting scheme is used in this work, where the local weight is denoted as $L_{jk} = log(f_{jk}) + 1$, f_{jk} is the frequency of occurrence of keyword t_k in document D_j. The global weight G_k is denoted as inverse document frequency as $G_k = log(M/M_k)$, for $i = (1, \cdots, , N)$, where M_k be the number of documents in which t_k is found and M is the total number of documents in the collection. Finally, the element w_{jk} is expressed as the product of local and global weight as $w_{jk} = L_{jk} * G_k$. This weighting scheme amplifies the influence of terms, which occur often in a document (e.g., *tf* factor), but relative rarely in the whole collection of documents (e.g., *idf* factor) [7]. A query D_q is also represented as a vector of length N as $\mathbf{f}_q^D = [\hat{w}_{q1}, \cdots, \hat{w}_{qi}, \cdots, \hat{w}_{qN}]^T$.

2.2   Concepts-Based Image Representation

In a heterogeneous collection of medical images, it is possible to identify specific local patches that are perceptually and/or semantically distinguishable, such as homogeneous texture patterns in grey level radiological images, differential color and texture structures in microscopic pathology and dermoscopic images, etc. The variation in these local patches can be effectively modeled as visual keywords by using supervised learning based classification techniques, such as

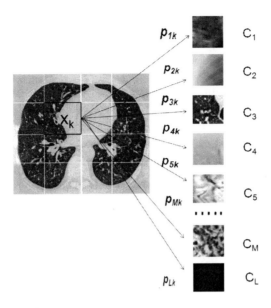

Fig. 4. Image encoding with probabilistic membership scores

the support vector machine (SVM) [8]. In its basic formulation, the SVM is a binary classification method that constructs a decision surface and maximizing the inter-class boundary between the samples. A number of methods have been proposed for multi-class classification by solving many two-class problems and combining their predictions.

In this research, we utilize a multi-class classification method by combining all pairwise comparisons of binary SVM classifiers, known as *one-against-one* or pairwise coupling (PWC) [9]. PWC constructs binary SVM's between all possible pairs of classes. Hence, for L classes, this method uses $L * (L - 1)/2$ binary classifiers that individually compute a partial decision for classifying a data point (image). During the testing of a feature $\mathbf{x}$, each of the $L * (L - 1)/2$ classifier votes for one class. The winning class is the one with the largest number of accumulated votes.

In order to perform the learning, a set of L labels are assigned as $C = \{c_1, \cdots, c_i, \cdots, c_L\}$, where each $c_i \in C$ characterizes a visual concept. The training set of the local patches that are generated by a fixed-partition based approach and represented by a combination of color and texture moment and edge histogram related features [10]. For SVM training, the initial input to the system is the feature vector set of the patches along with their manually assigned corresponding concept labels. Images in the data set are annotated with visual concept labels by fixed partitioning each image I_j into l regions as $\{\mathbf{x}_{1_j}, \cdots, \mathbf{x}_{k_j}, \cdots, \mathbf{x}_{l_j}\}$, where each $\mathbf{x}_{k_j} \in \Re^d$ is a combined color and texture feature vector. For each $\mathbf{x}_{k_j}$, the visual concept probabilities are determined by the prediction of the multi-class SVMs as [9]

$$p_{ik_j} = P(y = i \mid \mathbf{x}_{k_j}), \ 1 \le i \le L. \tag{1}$$

For example, Fig. 4 shows a particular region in a segmented image and its probabilistic membership scores to different local concept categories. Finally, the category label of x_{k_j} is determined as c_m, which is the label of the category with the maximum probability score. Hence, the entire image is thus represented as a two-dimensional index linked to the visual concept labels. Based on this encoding scheme, an image I_j is represented as a vector of visual concepts as

$$\mathbf{f}_j^I = [w_{j1}, \cdots, w_{ji}, \cdots w_{jL}]^{\mathrm{T}} \tag{2}$$

where each w_{ji} corresponds to the normalized frequency of a concept $c_i, 1 \le i \le L$ in image I_j. Here, the vector dimension equals to the number of local concept categories.

3   Multimodal Image Search

Let us consider q as a multi-modal query, which has an image part as I_q and a text part as D_q. The similarity between q and a multi-modal item j, which also has two parts (e.g., image (concept) I_j and text (context) D_j), is defined as

$$\mathrm{Sim}(q, j) = \omega_I \mathrm{Sim}_{\mathrm{concept}}(I_q, I_j) + \omega_D \mathrm{Sim}_{\mathrm{context}}(D_q, D_j) \tag{3}$$

Here, ω_I and ω_D are normalized inter-modality weights within the concept and context feature spaces, which subject to $0 \le \omega_I, \omega_D \le 1$ and $\omega_I + \omega_D = 1$. The effectiveness of the linear combination depends mainly on the choice of the modality weights, which can be found out experimentally.

In our multimodal framework, the individual image $\mathrm{Sim}_{\mathrm{concept}}(I_q, I_j)$ and text $\mathrm{Sim}_{\mathrm{context}}(D_q, D_j)$ based similarities are computed based on the Cosine distance measure [7]. In particular, similar documents (images) are expected to have small angles between their corresponding vectors. In many cases, the direction or angle of the vectors are a more reliable indication of the semantic similarities of the objects than the distance between the objects in the term-document space. Hence, to compare a query and document vector, the cosine similarity measure is applied as follows [7]

$$\mathrm{Sim}_{\mathrm{context}}(D_q, D_j) = \cos(\mathbf{f}_q^D, \mathbf{f}_j^D) = \frac{\sum_{i=1}^{N} w_{qi} * w_{ji}}{\sqrt{\sum_{i=1}^{N}(w_{qi})^2} * \sqrt{\sum_{i=1}^{N}(w_{ji})^2}} \tag{4}$$

where w_{qi} and w_{ji} are the weights of the term $\mathbf{t}_i$ in D_q and D_j respectively. In a similar way, cosine similarity measure is applied to the concept feature vector.

Due to the large number of images and vector size, it might take considerable amount of times to retrieve images from a collection. In the following section, we present a filtering approach based on multi-class classification on the multimodal input feature vector described earlier.

4 Modality Detection and Filtering

The variation of the medical image categories (e.g., modalities) at a global level can also be effectively modeled by the multi-class SVM as described in the previous section. For the SVM training, the input is a feature vector set of training images in which each image is manually annotated with a single modality label selected out of the M modalities. So, a set of M labels are defined as $\{\omega_1, \cdots, \omega_i, \cdots, \omega_M\}$, where each ω_i characterizes the representative image modality. In this context, given a multimodal feature vector $\mathbf{x}$, which is a simple concatenation of the context and concept feature vectors, the multi-class estimates the probability or confidence scores of each category as

$$p_m = P(y = \omega_m \mid \mathbf{x}), \text{for } 1 \leq m \leq M \tag{5}$$

The final category of a feature is determined based on the maximum probability score.

Algorithm 1. Multimodal Image Filtering

(Off-line): Select a set training images (docs) of M categories with associated category label for SVM learning. Perform SVM learning based on the input of the combined multimodal feature vector $[\mathbf{f}^D \cdot \mathbf{f}^I]$ for each training images (docs).

(Off-line): Predict the category of each database image by applying SVM and store the category vectors (Equation 6) of N database images as a category index along with the feature indexes.

(On-line): For a multimodal query image of parts I_q and D_q, determine the category vector as $\mathbf{p}_q = [p_{q_1}, p_{q_2}, \cdots, p_{q_M}]^{\mathrm{T}}$.

for $j = 1$ to N **do**

    Consider the top ranked $(n < M)$ category labels for I_q and I_j after sorting the elements in the category vectors.

    Construct the category label sets as S_q and S_j for the top ranked categories of I_q and I_j respectively. Here, $|S_q| = n$ and $|S_j| = n$.

    if $(S_q \cap S_j \neq \emptyset)$ **then**

        Consider $I(D)_j$ for further similarity matching (Equation 3)

    end if

end for

We finally utilize the information about category prediction of query and database images for image filtering to reduce the search space. The output of the above classification approach form a M-dimensional category vector of an image $I(D)$ as follows

$$\mathbf{p}_j = [p_{j_1}, \cdots, p_{j_m}, \cdots, p_{j_M}]^{\mathrm{T}} \tag{6}$$

Here, $p_{j_m}, 1 \leq m \leq M$, denotes the probability or class confidence score that an image $I(D)_j$ belongs to the category ω_m in terms of the multimodal feature vector.

During the off-line indexing process, this output is stored as the category vector of the database images in a *category index* along with the feature indices. Similar feature extraction and category prediction stages are performed on-line when the system is searched using an unknown query image. The category vector of a query image $I(D)_q$ and the vectors of the database images from the category index are evaluated to identify candidate target images in the collection, thereby filtering out irrelevant images from further consideration. To minimize misclassification errors, instead of only considering the image categories based on the highest obtained probability values, $n < M$ nearest classes of the target images to the query image are considered.

The process validates for class overlap between the query and target images. Generally, the value of $n << M$ to prevent inclusion of distant classes and provide effective filtering. A target image is only selected for further matching if at least one common category is found out between the top n categories of the query image and itself. This further reduces the risk of searching wrong images due to misclassification. Steps of the filtering algorithm are presented in Algorithm 1.

5   Experiments and Results

To evaluate the retrieval effectiveness, experiments are performed on the Image-CLEFmed 2010 benchmark medical image collection. The experimental results are generated based on the 16 ad hoc query topics (e.g., a short sentence or phrase describing the search request in a few words with one to three relevant images) that were initially generated based on a log file of Pubmed[4]. All topics were categorized with respect to the retrieval approaches expected to perform best, i.e., visual topics for CBIR, semantic topics for text retrieval and mixed topics for multi-modal retrieval. Each topic consisted of the query itself in three languages (English, German, French) and 2 to 3 example images for the visual part of the topic.

5.1   Training for SVM

A training set of about 2400 images provided by the ImageCLEFmed 2010 is used for SVM training for modality detection. The images are classified into one of the 8 modalities (e.g., CT, MR, XR, etc,) as shown in Table 1.

For the SVM training, we utilized the radial basis function (RBF) as kernel. A 10-fold cross-validation (CV) is conducted to find the best values of the tunable parameters C and γ of the RBF kernel as shown in Table 2.

For the visual concept generation based on the SVM learning, 30 local concept categories are manually defined, such as tissues of lung or brain of CT or MRI, bone of chest, hand, or knee x-ray, microscopic blood or muscle cells, dark or white background, etc. The training set consists of less then 1% images of the entire collection. Each image in the training set is partitioned into an 8×8 grid generating 64 non-overlapping regions, which is proved to be effective to generate the local patches. Only the regions that conform to at least 80% of a particular concept category are selected and labeled with the corresponding category label due to the consideration of robustness to noise [10]. After finding the best values

---

[4] http://www.pubmed.gov

Table 1. Image categories and number of training images

Modality	No. of Images
CT: Computerized tomography	314
GX: Graphics, typically drawing and graphs	355
MR: Magnetic resonance imaging	299
NM: Nuclear Medicine	204
PET: Positron emission tomography including PET/CT	285
PX: optical imaging (photographs, micrographs, gross pathology etc)	330
US: ultrasound including (color) Doppler	307
XR: x-ray including x-ray angiography	296

Table 2. 10-Fold Cross Validation (CV) Accuracy

Feature	C	γ	Accuracy
Concept	100	0.0002	73.89%
Context (Caption)	20	0.0002	90.50%
Context (Caption+Title+MeSH)	20	0.0002	90.54%
Combined (Context + Concept)	200	0.00001	95.39%

of the parameters $C = 200$ and $\gamma = 0.02$ of the RBF kernel with a 10-fold CV accuracy of 81.01%, they are utilized for the final training to generate the SVM model file. We utilized the *LIBSVM* software package [11] for implementing the multi-class SVM classifiers.

5.2 Performance Analysis

Results for different retrieval methods are computed using the latest version of TREC-EVAL[5] software based on the relevant sets of all topics, which were crated by the CLEF organizers by considering top retrieval results of all submitted runs of the participating groups. Results were evaluated using un interpolated (arithmetic) Mean Average Precisions (MAP) to test effectiveness, Geometric Mean Average Precision (GMAP) to test robustness, and Precision at rank 20 (P20).

It is clear from Table 3 that the best MAP score (0.1125) is achieved when a multimodal search is performed in a filtered image set. Although, we achieved a lower MAP score compared to the text only search approach when no filtering is applied based on multimodal search. This result might be an indication that the query topics are more semantic in nature and mixing with image features only lower the precision when search is performed on the entire collection. The other scores (e.g., GMAP, Rprec, and Bpref) also slightly improved when we compare filtering and without filtering approaches as shown in Table 3. Finally, from the

---

[5] http://trec.nist.gov/trec-eval/

Table 3. Retrieval Results based on the Query Topics (CLEF'10)

Feature	MAP	GMAP	Rprec	Bpref	P(20)
Concept	0.0010	0.0001	0.0049	0.0144	0.0063
Context	0.1058	0.0133	0.1261	0.1441	0.1906
Multimodal	0.0958	0.0133	0.1150	0.1605	0.1781
Multimodal (Filter)	0.1125	0.0159	0.1292	0.2176	0.1875

results, we can conjecture that the pre-filtering approach is indeed an effective one as the performances are always better when compared to the searched which were performed on the entire collection.

Further, an important benefit of searching on a filtered image set is gain in computation time. We tested the efficiency of the multimodal search scheme by comparing the average retrieval time for 16 query topics with and without applying the filtering scheme. The experiment was performed in an Intel Pentium Dual-Core CPU at 3.40 GHz with 3.5 GB of RAM running Microsoft Windows XP SP2 Professional operating system. The linear search time without filtering was twice as much as search on the filtered image set, suggesting that the proposed method is both effective and efficient.

6   Conclusions

In this paper, a novel framework for multi-modal interaction and integration is proposed for a diverse medical image collection with associated annotation of the case or lab reports. Unlike in many other approaches, where the search is performed with a single modality and without any classification information, we proposed to use the classification result directly in the retrieval loop and integrate the results obtained from both the text and imaging modalities. A standard image dataset has provided enough reliability for objective performance evaluation that demonstrates the efficacy of the proposed method.

Acknowledgment. This research is supported by the Intramural Research Program of the National Institutes of Health (NIH), National Library of Medicine (NLM), and Lister Hill National Center for Biomedical Communications (LH-NCBC). We thank the ImageCLEFmed [4] organizers for making the dataset available for the experiments.

References

1. Winfield, W., Lain, E., Horn, T., Hoskyn, J.: Eosinophilic cellulitislike reaction to subcutaneous etanercept injection. Arch. Dermatol. 142 (2), 218–220 (2006)
2. Müller, H., Michoux, N., Bandon, D., Geissbuhler, A.: A Review of Content-Based Image Retrieval Systems in Medical Applications Clinical Benefits and Future Directions. Int. J. of Med. Inform. 73 (1), 1–23 (2004)

3. Wong, T.C.: Medical Image Databases. Springer, New York (1998)
4. Müller, H., Kalpathy-Cramer, J., Eggel, I., Bedrick, S., Reisetter, J., Kahn. Jr., C.E., Hersh, W.R.: Overview of the CLEF, Medical Image Retrieval Track. In: CLEF (Notebook Papers/LABs/Workshops) (2010)
5. Lehmann, T.M., Güld, M.O., Deselaers, T., Keysers, D., Schubert, H., Spitzer, K., Ney, H., Wein, B.B.: Automatic categorization of medical images for content-based retrieval and data mining. Comput. Med. Imag. and Graph. 29, 143–155 (2005)
6. Florea, F., Müller, H., Rogozan, A., Geissbuhler, A., Darmoni, S.: Medical image categorization with MedIC and MedGIFT. In: Proc. Med. Inform. Europe (MIE 2006), Maastricht, Netherlands, pp. 3–11 (2006)
7. Yates, R.B., Neto, B.R.: Modern Information Retrieval. Addison-Wesley (1999)
8. Vapnik, V.: Statistical Learning Theory. Wiley, New York (1998)
9. Wu, T.F., Lin, C.J., Weng, R.C.: Probability Estimates for Multi-class Classification by Pairwise Coupling. J. of Mach. Learn. Research 5, 975–1005 (2004)
10. Rahman, M.M., Antani, S.K., Thoma, G.R.: A Medical Image Retrieval Framework in Correlation Enhanced Visual Concept Feature Space. In: Proc. 22nd IEEE International Symposium on Computer-Based Medical Systems (CBMS), Albuquerque, New, Mexico, USA, August 3-4 (2009)
11. Chang, C.C., Lin, C.J.: LIBSVM: a library for support vector machines. Software (2001), http://www.csie.ntu.edu.tw/cjlin/libsvm

Using MeSH to Expand Queries
in Medical Image Retrieval

Jacinto Mata, Mariano Crespo, and Manuel J. Maña

Dpto. de Tecnologías de la Información. Universidad de Huelva
Ctra. Huelva - Palos de la Frontera s/n. 21819 La Rábida, Huelva
{jacinto.mata,mariano.crespo,manuel.mana}@dti.uhu.es

Abstract. The presence of huge collections of medical images in scientific repositories and hospital databases has given rise to increasing interest in access to this information. This paper addresses the issue, focusing on image retrieval based on textual information related to the image. The initial hypothesis is that query expansion could improve the effectiveness of image retrieval systems. In this proposal, several information elements contained in MeSH ontology were used. The ImageCLEF 2009 and 2010 document collections were used for the experiment. Results showed a slight increase in MAP and a more significant difference when the evaluation was performed using the F-measure in 2009 collection. The final conclusion is that query expansion is not sufficient to achieve a substantial improvement in the efficacy of this type of information retrieval systems.

Keywords: Text-based image retrieval, medical domain, query expansion, ontologies.

1   Introduction

Nowadays there is a large amount of biomedical information in electronic format. There are many huge collections of medical data with visual and textual information available to researchers, health professionals and all those interested in this type of information in general. Information retrieval systems have traditionally focused their efforts on improving the accessibility of textual information. However, there is growing interest in optimising access to visual information. Image retrieval is an area less explored than text retrieval. There are basically two approaches to address the image retrieval task: the approach based on the visual content of the image (content-based image retrieval) and the approach based on textual information about the image (text-based image retrieval). This paper presents a work based on the textual approach.

One of the main problems in image retrieval when taking the associated textual information is the difficulty of making a correct annotation of image content. In many situations it is difficult to express what happens in an image (actions, feelings...). Moreover, natural language has many complications (synonymy, hyperonymy, hyponymy, use of abbreviations, etc.).

In most cases, the use of external resources greatly enhances the performance of information retrieval systems. Great research efforts are currently underway to improve

H. Müller et al. (Eds.): MCBR-CDS 2011, LNCS 7075, pp. 36–46, 2012.
© Springer-Verlag Berlin Heidelberg 2012

these resources, especially in the biomedical field, to help the end user handle these large volumes of information [1, 2]. Ontologies are perhaps the most widely used resource. In the field of biomedicine there are a large number of thesauri and ontologies. Some of the most commonly used are GO[1] [3], MeSH[2] [4] and UMLS[3] [1].

In this paper, a survey on the use of MeSH ontology for query expansion to improve a medical image retrieval system is presented. For the survey and experimentation we employed the collections used in the medical image retrieval task (ImageCLEF[4]) in 2009 [5] and 2010 [6]. In these collections, the information linked to each image (metadata) consists of the image caption and the title of the article to which it belongs. It also includes the ID of the article to access the full text.

The rest of the paper is organised as follows. Section 2 describes the most relevant related work. Section 3 presents the collection of documents used in the survey and experimentation. Section 4 describes the expansion strategies. In Section 5 the results obtained in the experiments are shown and discussed. Finally, conclusions and future works are outlined in Section 6.

2   Related Work

Today there are several image retrieval systems. Most of them are specially designed to work in the biomedical domain. Some examples of these systems are *Yale Image Finder* [7], *ARRS Goldminer* [8] and *BioText* [9].

These systems obtain the images using the title text of the documents, the figure captions, abstracts or the whole document. However, due to the characteristics of the collections, in many cases these systems become less effective and therefore do not provide the expected results.

There are several works where studies on the effect of the use of the MeSH and UMLS ontologies for query expansion are presented. In [10], the authors investigate a query expansion strategy process using an advanced PubMed[5] search called *Automatic Term Mapping* (ATM). To perform the study they used a collection of 64 queries and around 160,000 citations from MEDLINE that were used in the TREC Genomics Track for the years 2006 and 2007. Results indicated an increase in F-measure of 21.5% and 23.3% in the collections for 2006 and 2007 respectively, using query expansion. The authors concluded that the query expansion using MeSH in PubMed can improve the recovery efficacy.

In [11], the authors showed their experiments in the medical information retrieval task in the ImageCLEF competition in the year 2008. Three different types of collection were created. For the first collection, the image captions and titles of the documents were used; the second collection was done with the captions, titles and text of the section referencing images, while the third collection consisted of the full text. The query expansion was carried out with the MeSH and UMLS ontologies, obtaining the

---

[1] http://www.geneontology.org/
[2] http://www.nlm.nih.gov/mesh/meshhome.html
[3] http://www.nlm.nih.gov/research/umls/
[4] http://www.imageclef.org/
[5] http://www.ncbi.nlm.nih.gov/pubmed/

best results with query expansion using MeSH and indexing the captions and titles of the documents. An improvement of 12.5%was reached for the MAP compared to baseline.

The same authors [12] showed the results obtained by query expansion using the *entry terms* in MeSH ontology. They used the collections of the medical information retrieval task of the ImageCLEF in 2005 and 2006. In that work, they used textual information and visual information and concluded that there was an improvement in the retrieval efficiency using both strategies together. Specifically, they achieved an improvement in the MAP measure of 25.07% and 37.3% compared with their baseline in the collections for 2006 and 2005, respectively.

The main difference between previous works and the proposal presented in this paper is basically the selection of the information used for query expansion. Our goal is to explore some of the links provided by these ontologies in order to identify a set of strategies for query expansion.

3   Collection Description

The collections of images selected to carry out the system evaluation were provided by the ImageCLEF organisation in 2009 and 2010. The ImageCLEF 2009 collection consists of 74,902 images and 25 queries (named topics, in ImageCLEF) while the 2010 collection is composed of 77,495 images and 16 topics. The participation was very broad in both years: 38 research groups around the world in 2009 and 51 in 2010.

ImageCLEF provides participants in the competition two files in XML format with the collections of documents and queries (topics). For system evaluation there is a third text file with relevance judgments for each query image (ground truth).

The XML file image collection consists of one record for each image. The metadata for each record are composed of the image identifier, its URL, the image caption, the title of the paper in which the image appears, the PubMedpaper ID, the paper URL and the name of the image file. Figure 1 shows an example of this kind of record.

```
<record>
<figureID>27979</figureID>
<figureURL>http://radiology.rsnajnls.org/cgi/content/full/210/1/11/F1</figureURL>
<caption> Figure 1.Illustration of a neonate at autopsy whose demise was attributed to
"thymic death." The caption drew attention to the "enormous size of the thymus," which
is actually normal in appearance. (Reprinted, with permission, from reference 6.)
</caption>
<title>The right place at the wrong time: historical perspective of the relation of the
thymus gland and pediatric radiology
</title>
<pmid>9885579</pmid>
<articleURL>http://radiology.rsnajnls.org/cgi/content/full/210/1/11</articleURL>
<imageLocalName>27979.jpg</imageLocalName>
</record>
```

Fig. 1. Example of ImageCLEFimage record

The judgment file consists of four columns. The first column specifies the query number. The third stands for the imager identifier. The fourth column shows the relevance (a binary value, where 0 is not relevant and 1 is relevant). The second column is ignored for this kind of evaluation. Figure 2 shows the file judgment format.

1	0	227775	0
1	0	227776	0
1	0	198482	0
....			
2	0	129252	1
2	0	53729	0
2	0	50647	1
....			
25	0	66087	0
25	0	187404	0
25	0	125313	0

Fig. 2. Fragment of ImageCLEF judgment file

The information provided for each topic comprises the query identifier, the topic type (visual, semantic and mixed) and the English topic text and its French and German translations. For this work, only the information in English was used. Figure 3 shows a topic record example.

```
<topic>
    <ID>2</ID>
    <TYPE>visual</TYPE>
    <EN_DESCRIPTION>
            Breast cancer mammogram
    </EN_DESCRIPTION>
    <FR_DESCRIPTION>
            Mammographies d'un cancer du sein
    </FR_DESCRIPTION>
    <DE_DESCRIPTION>
            MammogrammmitBrustkrebs
    </DE_DESCRIPTION>
</topic>
```

Fig. 3. Example of ImageCLEF topic record

4   Query Expansion Using MeSH

The query expansion term is used in a search engine when new terms are added to the user's query in order to increase efficiency in the recovery. Recently, systems based on query expansion are significantly improving their results, making use of external resources such as ontologies and lexical hierarchies.

Ontologies usually represent the knowledge of a specific domain as a set of concepts and relationships between them. In the biomedical domain, many terminological and ontological resources are available. There are also several Natural Language Processing (NLP) tasks where ontologies are used: information retrieval, question answering, automatic summarising or classification, among others. However, due to the large amount of information they provide, in most cases it is not easy to use them in IR tasks. In [13], the authors proposed an algorithm for refining ontologies for information retrieval tasks.

In the experiments presented in this paper, the MeSH initiative from the *National Library of Medicine* was used. MeSH is a controlled vocabulary used for indexing articles from Medline. It consists of sets of terms called descriptors, arranged in a hierarchical structure that enables the search at different levels of specificity. There are currently 26,142 MeSH descriptors or *Main Headings*. There are also over 177,000 alternative expressions, synonyms and terms related to these descriptors, named *entry terms*.

MeSH ontology offers many possibilities for expanding the query terms. In this paper, we propose some strategies for expansion based on the *entry terms* similar to those used in [12]. Several strategies based on other elements offered by the ontology are also presented.

Many times a descriptor is made up of more than one term. For example, if the query *Pituitary Adenoma* was made for each term independently, the *Pituitary* term does not correspond with any descriptor. However, the union of the two terms corresponds to a descriptor itself as "*Pituitary Adenoma*", which is a biomedical concept.

That is the reason why each query was pre-processed by dividing it into n-grams, with the aim of exploring all the possibilities offered by the query to obtain sequences that are MeSH descriptors and *entry terms*. Below is an example of processing a query with n-grams.

Query: Breast cancer mammogram

N - Grams
>   (1): Breast
>   (2): Breast cancer
>   (3): Breast cancer mammogram
>   (4): cancer
>   (5): cancer mammogram
>   (6): mammogram

Where the n-gram 2 and 4 are *entry terms* and 1 is a descriptor.

The following sections describe the strategies used to study expansion and experimentation.

4.1   Cross-Referencing Based Techniques

Specifically, two types of cross-references (*SeeRelatedDescriptor* and *ConsiderAlso*) and the hierarchical structure in which MeSH organises its own descriptors were used. The combinations that are descriptors in MeSH are expanded with the content of the element *SeeRelatedDescritor* or *ConsiderAlso*. The first one associates the descriptor with other descriptors related through a cross-reference. The aim of such partnerships is to provide other descriptors that may be more appropriate for a particular case. For example, the term *cancer* is expanded, among others, with the terms *Carcinogens, Antibodies, Neoplasm, Genes* and *Tumour Suppressor*.

ConsiderAlso is an element of reference to other descriptors having related linguistic roots. For example, the term *brain* references the linguistic roots *BRAIN-* and *ENCEPHAL-*.

The third expansion strategy is based on the tree structure whereby MeSH organises its descriptors. In this case, if the descriptor is a parent node, it is expanded with its child descriptors. If the descriptor does not have any children there is no expansion. Figure 4 shows a brief MeSH tree excerpt which indicates that the *Brain* descriptor has seven children while the *Central Nervous System* descriptor has three.

Nervous System [A08]
    Central Nervous System [A08.186]
        ▶ Brain [A08.186.211]
                Blood-Brain Barrier [A08.186.211.035]
                Brain Stem [A08.186.211.132] +
                Cerebral Ventricles [A08.186.211.276] +
                Limbic System [A08.186.211.464] +
                Mesencephalon [A08.186.211.653] +
                Prosencephalon [A08.186.211.730] +
                Rhombencephalon [A08.186.211.865] +
            Meninges [A08.186.566] +
            Spinal Cord [A08.186.854] +

Fig. 4. Excerpt from MeSH Tree

4.2   Techniques Based in Entry Terms

The first expansion strategy consists of exploring the MeSH tree by checking if the query n-gram is a descriptor. If the n-gram is a descriptor, the query is expanded using all the *entry terms* of the descriptor. If the n-gram is not a descriptor, we check if it is an *entry term*. If so, the descriptor and all its entry terms are added to the expansion.

The second strategy has only a small variation from the first. When a n-gram in the query is a descriptor, the query is expanded with the *entry terms* of the *preferred concept*, instead of all the *entry terms* of that descriptor.

When the results of these expansion strategies were calculated, it was found that they introduced too much noise into the queries and the results were not as good as expected. To this end, a filtering of the query was carried out. With this process, two new strategies for expansion were designed, one of them stricter (closed strategy) and the other less strict (open strategy). Figure 5 shows an example of a filtering process.

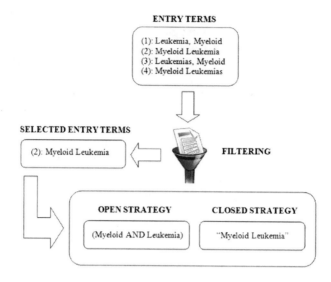

Fig. 5. Example of a filtering process

5   Results and Discussion

This section details the experiments that were conducted to assess various expansion strategies. Usually, in ImageCLEF competitions, the most relevant measure in the organisation was the *Mean Average Precision* (MAP). This is a measure of a single value which takes into account the position in the ranking of relevant documents retrieved. It is calculated as the average of the accuracies obtained each time a relevant document is retrieved. For a collection of information needs, it is averaged as shown in (1).

$$MAP = \frac{1}{m} \sum_{j=1}^{m} \frac{1}{r_j} \sum_{k=1}^{r_j} \text{Precisión}(R_k) \tag{1}$$

Where:

    m: Query number.
    r_j: Number of relevant documents in the query j.
    R_k: k-th relevant document retrieved, of the ranking of documents retrieved.

For the system evaluation, TREC_EVAL software developed by the TREC (Test REtrieval Conference) was used. This software calculates the measures most widely used to assess retrieval efficacy. To evaluate the influence of the expansion of the queries on the retrieval result, a baseline was designed. Two indexes were drawn up, one with the captions of the 2009 collection and another with the captions of the 2010 collection. The Lucene search engine was used for this process.

The following sections shows the results obtained with the expansion strategies described in the previous section. The improvements achieved with the proposed expansion techniques are lower than expected. However, these results are not comparable to previous works [11] and [12] because the collections used in the experiments are different.

Table 1. Results from MAP measure for each query

COLLECTION 2009							
Q	BL	SRD	CA	CT	ETCS	ETOS	ETPCCS	ETPCOS
1	0,1064	0,1064	0,1064	**0,1404**	0,1064	0,1064	0,1064	0,1064
2	**0,3278**	0,2020	0,2001	0,1632	0,1225	0,1046	0,1225	0,1046
3	**0,3311**	**0,3311**	**0,3311**	**0,3311**	0,3008	0,3019	0,3008	0,3019
4	0,2238	0,2649	0,2238	**0,3208**	0,2213	0,2213	0,2213	0,2213
5	0,4644	0,4644	0,3492	**0,5295**	0,3494	0,3494	0,3494	0,3494
6	0,3231	0,3168	0,3231	**0,3379**	0,2935	0,3119	0,2935	0,3119
7	**0,3062**	**0,3062**	0,3014	**0,3062**	0,3038	0,3038	0,3038	0,3038
8	0,5794	0,5794	0,5794	**0,5986**	0,5794	0,5794	0,5794	0,5794
9	**0,3347**	0,2415	0,1937	**0,3347**	0,2389	0,2218	0,2564	0,2496
10	0,3010	0,2694	0,3010	0,3010	**0,3662**	0,3301	**0,3662**	0,3301
11	**0,4646**	0,2515	**0,4646**	0,3985	**0,4646**	**0,4646**	**0,4646**	**0,4646**
12	0,4777	0,4777	0,4777	**0,5269**	0,4777	0,4777	0,4777	0,4777
13	0,0185	0,0018	0,0009	0,0043	**0,0314**	**0,0314**	**0,0314**	**0,0314**
14	0,7629	0,7629	0,7629	0,8620	**0,8649**	0,7605	0,6945	0,3060
15	0,5257	0,5257	0,5257	0,5257	0,5105	0,5117	**0,5261**	0,5249
16	0,1652	0,1651	0,1652	0,1652	**0,1808**	**0,1808**	0,1652	0,1713
17	**0,0168**	0,0109	0,0085	**0,0168**	**0,0168**	**0,0168**	**0,0168**	**0,0168**
18	0,7178	0,6602	0,7178	0,7178	0,6966	0,7198	0,7045	**0,7257**
19	**0,6108**	**0,6108**	**0,6108**	**0,6108**	0,5960	0,5301	0,5960	0,5301
20	**0,1320**	**0,1320**	**0,1320**	0,0887	0,0883	0,0803	0,0884	0,0822
21	**0,1217**	**0,1217**	**0,1217**	**0,1217**	0,1111	0,079	0,1111	0,0790
22	0,3591	0,1823	0,3443	0,3814	0,3152	0,4091	0,3936	**0,4145**
23	0,1038	0,1038	0,1038	**0,2877**	0,1038	0,1038	0,1038	0,1038
24	0,3720	0,3720	0,3720	0,4001	0,1581	0,1904	0,3632	**0,4103**
25	0,3331	0,3331	0,3331	0,3331	**0,3242**	0,3182	**0,3242**	0,3182
MV	0,3391	0,3117	0,3220	**0,3522**	0,3129	0,3082	0,3184	0,3006
COLLECTION 2010							
Q	BL	SRD	CA	CT	ETCS	ETOS	ETPCCS	ETPCOS
1	0,3669	0,3669	0,3669	0,3928	**0,3669**	**0,3669**	0,3669	**0,3669**
2	0	0	0	0	0	0	0	0
3	0,2042	0,2042	0,2042	0,2042	0,2627	**0,2627**	0,2626	0,2626
4	0,0356	0,0372	**0,1450**	0,0431	0,0275	0,0275	0,0473	0,0528
5	**0,2983**	0,2696	**0,2983**	**0,2983**	0,2976	0,2807	0,2976	0,2807
6	**0,3239**	0,1748	0,0832	0,3167	0,3104	0,2354	0,3163	0,311
7	**0,5011**	**0,5011**	**0,5011**	**0,5011**	**0,5011**	**0,5011**	**0,5011**	**0,5011**
8	**0,3333**	**0,3333**	**0,3333**	0,0294	0,0833	0,0714	0,2500	0,1429
9	**0,6435**	0,6209	**0,6435**	**0,6435**	0,6209	0,4055	0,6353	0,6208
10	**0,5728**	0,1490	0,2652	0,3969	0,4631	0,4608	0,4840	0,4919
11	0,0652	0,0337	0,0069	0,0576	0,0425	**0,1008**	0,0117	0,0082
12	0,2561	0,2561	0,2561	0,2561	0,2561	0,2561	**0,2733**	**0,2733**
13	**0,1563**	**0,1563**	**0,1563**	**0,1563**	**0,1563**	**0,1563**	**0,1563**	**0,1563**
14	**0,7157**	**0,7157**	**0,7157**	**0,7157**	**0,7157**	**0,7157**	**0,7157**	**0,7157**
15	**0,5851**	0,3550	**0,5851**	**0,5851**	0,5398	0,4721	0,5398	0,4721
16	0	0	0	0	**0,0105**	0,0099	**0,0105**	0,0099
MV	**0,3161**	0,2609	0,2851	0,2873	0,2909	0,2702	0,3043	0,2916

Table 1 shows the MAP measure obtained by all the queries with each of the expansion strategies, where:

- Q: Query number.
- BL: Baseline.
- SRD: SeeRelatedDescriptor.
- CA: ConsiderAlso.
- CT: MeSH Tree with concepts.
- ETCS: Entry Terms Closed Strategy
- ETOS: Entry Terms Open Strategy
- ETPCCS: Entry Terms Preferred Concept Closed Strategy
- ETPCOS: Entry Terms Preferred Concept Open Strategy

The expansion technique which provided best results in the 2009 collection was achieved by expanding from the MeSH tree (CT), obtaining an average value for the MAP of 0.3522, representing an increase of 4% compared to our baseline. On the other hand, the best strategy in 2010 was ETPCCS, although neither of them improved the value of the baseline.

The last row shows the mean values (MV) for the baseline and the 7 strategies used. Note that in the 2009 edition, the MAP of the winner system was 0.43 and that a measure of 0.35 would have been placed among the top 12 runs. In addition, in the 2010 edition, the winner system obtained a MAP of 0.338 and our baseline was 0.3161. With this value, our system would have been placed among the top 6 runs.

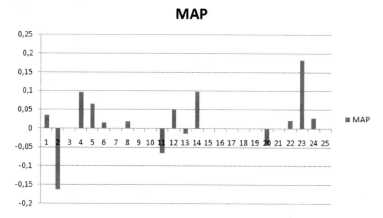

Fig. 6. Histogram for the 25 queries using AC expansion strategy (collection 2009)

Figure 6 shows the histogram for each query of the year 2009. It can be seen that 10 of the 25 queries improved on this measure while in 4 of them the MAP decreased, and 11 remained unchanged because there was no expansion. This fact suggests that it would be desirable do an alternative expansion in order to have a choice when there is no expansion with this strategy. It is worth noting as significant the decrease that occurs in query 2, which significantly lowers the average MAP value.

Table 2. Average value of precision, recall and F measure with each expansion strategy

Measure	BL	SRD	CA	CT	ETCS	ETOS	ETPCCS	ETPCOS
COLLECTION 2009								
Measure	**BL**	**SRD**	**CA**	**CT**	**ETCS**	**ETOS**	**ETPCCS**	**ETPCOS**
Precision	**0,1051**	0,1035	0,1048	0,0899	0,1032	0,1008	0,1026	0,1003
Recall	0,6801	0,6611	0,6700	**0,6875**	0,6805	0,6735	0,6696	0,6649
F	0,1185	0,1160	0,1182	**0,1319**	0,1160	0,1126	0,1150	0,1117
COLLECTION 2010								
Measure	**BL**	**SRD**	**CA**	**CT**	**ETCS**	**ETOS**	**ETPCCS**	**ETPCOS**
Precision	0,0447	0,0414	0,0443	0,0452	0,0451	**0,0455**	0,0451	0,0451
Recall	0,6257	0,6185	0,6388	0,6443	0,6442	**0,6476**	0,6442	0,6442
F	0,0736	0,0690	0,0730	0,0746	0,0745	**0,0752**	0,0745	0,0745

Although in ImageCLEF the measure used to evaluate the results was the MAP, in this work, a comprehensive study of the expansion techniques developed was carried out. In addition, the precision, recall and F-measure were calculated. Table 2 shows the values obtained with each of the expansion strategies.

As can be seen, the technique that achieves the best result as F-measure in the collection of 2009 is also the MeSH tree (CT). In this case, the increase was 11.3% compared to our baseline. In the 2010 collection, the best results are reached using the ETOS strategy. However the increase was not very significant (2.1% for ETOS strategy and 1.3% for CT strategy).

6   Conclusions and Future Work

In this paper we have presented a study on the use of different query expansion strategies using one of the most widely used ontologies in the medical domain, with the aim of enhancing the efficacy of a textual content-based image retrieval system.

Different elements of MeSH ontology were chosen for expansion. The results of our experiments showed that not all expansion strategies are able to improve the effectiveness of the system. However, it was proven that MeSH hierarchical structure could be used to achieve a slight improvement in both MAP and F-measure. This improvement was achieved in the 2009 collection.

This study verified the difficulty of finding an appropriate strategy for query expansion. We believe that there are information elements or element combinations in MeSH that might be used to expand the queries and could substantially improve an image retrieval system.

Future work will continue research into other query expansion strategies and the use of other ontologies, such as UMLS. We shall also use textual information associated with each image. In this regard, we intend to add sections to the index where the images are described in the paper along with the captions. In this work, only queries were expanded. In future experiments, we shall explore the possibilities of concept indexing using UMLS.

Acknowledgments. This work was partially funded by the Spanish Ministry of Science and Innovation, the Spanish Government Plan E and the European Union through ERDF (TIN2009-14057-C03-03).

References

1. Bodenreider, O.: The Unified Medical Language System (UMLS): integrating biomedical terminology. Nucleic Acids Research 32, 267–270 (2004)
2. Nelson, S.J., Johnston, D., Humphreys, B.L.: Relationships in medical subject headings. Relationships in the Organization of Knowledge, pp. 171–184. Kluwer Academic Publishers (2001)
3. Stevens, R., Goble, C.A., Bechhofer, S.: Ontology-based knowledge representation for bioinformatics. Brief Bioinformatics 1(4), 398–414 (2000)
4. Nelson, S.J., Schopen, M., Savage, A.G., Schulman, J.L., Arluk, N.: The MeSH translation maintenance system: structure, interface, design and implementation. In: Fieschi, M., et al. (eds.) Proceedings of the 11th World Congress on Medical Informatics, pp. 67–69 (2004)
5. Müller, H., Kalpathy–Cramer, J., Eggel, I., Bedrick, S., Radhouani, S., Bakke, B., Kahn Jr., C.E., Hersh, W.: Overview of the CLEF 2009 Medical Image Retrieval Track. In: Peters, C., Caputo, B., Gonzalo, J., Jones, G.J.F., Kalpathy-Cramer, J., Müller, H., Tsikrika, T. (eds.) CLEF 2009. LNCS, vol. 6242, pp. 72–84. Springer, Heidelberg (2010)
6. Müller, H., Kalpathy–Cramer, J., Eggel, I., Bedrick, S., Reisetter, J., Kahn Jr., C.E., Hersh, W.: Overview of the CLEF 2010 medical image retrieval track (2010),
http://www.clef2010.org/resources/proceedings/
ImageCLEF2010_medOverview.pdf
7. Xu, S., McCusker, J., Krauthammer, M.: Yale Image Finder (YIF): a new search engine for retrieving biomedical images. Bioinformatics 24(17), 1968–1970 (2008)
8. Kahn Jr., C.H., Thao, C.: GoldMiner: A Radiology Image Search Engine. American Journal of Roentgenology 188, 1475–1478 (2007)
9. Hearst, M., Divoli, A., Guturu, H., Ksikes, A., Nakov, P., Wooldridge, M.A., Ye, J.: BioText Search Engine: beyond abstract search. Bioinformatics 23(16), 2196–2197 (2007)
10. Lu, Z., Kim, W., Wilbur, W.: Evaluation of query expansion using MeSH in PubMed. Information Retrieval 12(1), 69–80 (2009)
11. Díaz-Galiano, M.C., García-Cumbreras, M.A., Martín-Valdivia, M.T., Ureña-López, L.A., Montejo-Ráez, A.: Query Expansion on Medical Image Retrieval: MeSH vs. UMLS. In: Peters, C., Deselaers, T., Ferro, N., Gonzalo, J., Jones, G.J.F., Kurimo, M., Mandl, T., Peñas, A., Petras, V. (eds.) CLEF 2008. LNCS, vol. 5706, pp. 732–735. Springer, Heidelberg (2009)
12. Díaz, M.C., Martín, M.T., Ureña, L.A.: Query expansion with a medical ontology to improve a multimodal information retrieval. Computers in Biology and Medicine 4, 396–403 (2009)
13. Jimeno, A., Berlanga, R., Rebholz, D.: Ontology refinement for improved information retrieval. Information Processing & Management 46(4), 426–435 (2010)

Building Implicit Dictionaries Based on Extreme Random Clustering for Modality Recognition

Olivier Pauly, Diana Mateus, and Nassir Navab

Computer Aided Medical Procedures,
Technische Universität München, Germany
{pauly,mateus,navab}@cs.tum.edu

Abstract. Introduced as a new subtask of the ImageCLEF 2010 challenge, we aim at recognizing the modality of a medical image based on its content only. Therefore, we propose to rely on a representation of images in terms of words from a visual dictionary. To this end, we introduce a very fast approach that allows the learning of implicit dictionaries which permits the construction of compact and discriminative bag of visual words. Instead of a unique computationally expensive clustering to create the dictionary, we propose a multiple random partitioning method based on Extreme Random Subspace Projection Ferns. By concatenating these multiple partitions, we can very efficiently create an implicit global quantization of the feature space and build a dictionary of visual words. Taking advantages of extreme randomization, our approach achieves very good speed performance on a real medical database, and this for a better accuracy than K-means clustering.

1 Introduction

With the goal of promoting multi-modal information retrieval, ImageCLEF proposes each year a medical retrieval challenge [10]. Made accessible by the Radiological Society of North America (RSNA), the database for this challenge contains more than seventy thousands images taken from publications that appeared in the journals Radiology and Radiographics. Consisting of 2D images in JPEG format, this collection counts medical images from different modalities, but also photographs, drawings and graphics. Moreover, many have been "processed" *e.g.* zoomed, cropped or annotated by medical experts. Because of their high variability, retrieval of medical images in such multi-modal database is a challenging task. In [7], Kalpathy-Cramer *et al.* demonstrated the importance of recognizing first the modality of images in order to improve the precision of image retrieval. Motivated by this, a new subtask was organized last year at the ImageCLEF 2010 challenge. In this paper, we propose to tackle the problem of recognizing the modality of an image based on its visual content only. Since similar anatomies appear in the different classes, we can not rely on semantic information to discriminate the modalities. However, since each imaging system is based on a different physical phenomenon, resulting images show particular local visual signatures such as textural and noise patterns at small scales. For instance, ultrasound images contain particular speckle patterns while relevant information is contained in low fequency edges. Hence, to recognize modalities independently of organs or anatomical structures

H. Müller et al. (Eds.): MCBR-CDS 2011, LNCS 7075, pp. 47–57, 2012.

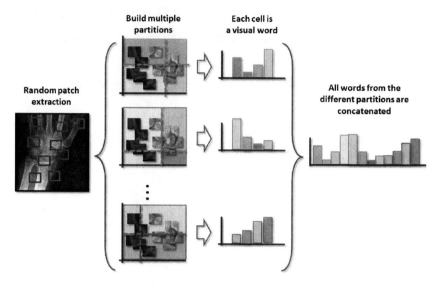

Fig. 1. Dictionary Learning Overview: First, visual features are extracted at random positions in the images. Then, multiple independent partitions of the feature space are built. Finally, each cell of these partitions are associated to a visual words.

appearing in the images, we propose to rely on local textural and noise patterns information extracted at random positions of the image. To efficiently represent the global statistics of appearance of such local signatures in an image, a bag of visual words (BoW) can be constructed based on a visual dictionary. While K-means clustering is a classical approach to build visual dictionaries, it suffers from several limitations such as its computational cost and its dependence on the quality of its initialization.

In this paper, our main technical contribution is an extreme random clustering approach to build efficiently implicit dictionaries and construct discriminative BoWs. As shown on Fig.1, our approach begins with the random sampling of the input feature space by extracting low level visual features at random positions in the images. Then, multiple random partitions are built using Extreme Random Subspace Projection Ferns. Finally, each cell of these partitions is associated to a visual word to form what we call an implicit dictionary. Experiments conducted on CT, MR, PET, US and X-ray images taken from the ImageCLEF 2010 database show that our approach is a fast alternative to K-means clustering which provides better performance in terms of accuracy and speed.

2   Related Work

In computer vision, bag of visual words (BoW) have become a standard representation tool for multi-class image recognition tasks. BoWs describe the content of an image in terms of the frequency of appearance of the so called visual words. The extraction of these visual words relies on a quantization of the high dimensional space spanned by low-level visual features [19]. A classical approach to quantize the feature space is K-means clustering [17]. While K-means is highly dependent on the quality of its initialization, its major drawback is its computational cost. Indeed, performing several runs of K-means clustering in a high dimensional space may last a few days, which is not suitable for updating the BoW representation of a medical database that changes on a daily basis.

During the last decade, efforts have been made to overcome the limitations of K-means clustering to learn dictionaries. For instance, Nister *et al.* [11] introduced a tree-based approach for CD-cover recognition relying on hierarchical K-means that shows better performance in terms of speed. Using mean shift, Jurie *et al.* [6] overcame K-means' tendency to draw cluster centers towards denser regions of the feature space. Winn *et al.* [18] generate compact dictionaries by merging some of the visual words, and this, without loss of discriminativity. In order to improve the discriminativity of dictionaries, Perronin *et al.* [15] introduced an approach that combines universal and class specific dictionaries and uses generative models. Yang *et al.* [19] proposed instead to unify the unsupervised clustering with the training of a classifier. In the same direction, Mairal *et al.*[8] presented an optimization framework to learn simultaneously a sparse representation from a dictionary and its associated classifier.

As the papers cited above, we aim at improving the learning of dictionaries. The focus of the proposed method differs from those in [15, 19, 8], as we propose to replace a single but complex clustering step by the construction of multiple random partitions to very efficiently quantize the feature space and thereby learn an implicit dictionary.

Since trees are able to identify natural clusters in high-dimensional spaces, random forests have been recently applied to dictionary learning. For instance, Moosmann *et al.* [9] and later Shotton *et al.* [16] proposed to build visual dictionaries for object categorization by using leaves and nodes from all trees of a random forest as visual words. Such dictionaries have shown state of the art performances while benefiting from fast learning and evaluation. To identify clusters with a higher resolution, Perbet *et al.* proposed in [14] to compute the intersections of all partitions and to represent them with the nodes of a graph, which is then clustered with a Markov Cluster algorithm. If this method leads to an explicit global partition of the feature space, its construction requires to solve a second clustering problem. Moreover, reducing the redundancy of the dictionary may lead to a loss of discriminativity. Following the idea of using leaves and nodes as visual words from Moosmann *et al.* [9], we propose a very efficient dictionary learning approach based on extreme randomized clustering using ferns we call ***Extreme Random Subspace Projection Ferns***, which provides a compact structure and benefits from very fast training and evaluation.

3 Proposed Method

We formulate the imaging modality recognition problem as an instance of a multi-class classification problem. Our contribution is a method to learn a visual dictionary based on extreme randomization in order to construct bag of visual words (BoWs) to represent the images we want to classify. With an extreme random partitioning algorithm we call ***Extreme Random Subspace Projection Ferns*** (ERSP), we can very efficiently construct multiple quantizations of the feature space that we then use to build an implicit dictionary. Our approach consists in the following steps:

1. **Extract** random points from the visual feature space.
2. **Build** efficiently multiple random partitions of the feature space with ERSP.
3. **Concatenate** these multiple random partitions.
4. **Associate** each cell to a visual word to construct an implicit dictionary.
5. **Build** bag of visual words (BoWs).
6. **Classify** using SVM with RBF kernel [2].

In contrast to K-means clustering, our approach is very fast and efficient, it is neither dependent on initialization nor requires the number of clusters to be known beforehand. Moreover, introducing randomization in the clustering phase permits to gain independence of the available training set, which in turn provides better generalization in the case of undersampled feature space, unbalanced data or noisy labeling. Finally, the proposed method can be used in a supervised as well as semi-supervised setting. Next, we describe the steps of the method enumerated above in details.

3.1   Visual Feature Space

The choice of suitable low-level visual features is crucial. In classical object recognition, features are especially designed for recognizing an object subject to different imaging conditions. Recognizing imaging modalities contrasts from classical recognition since similar objects may appear in several classes, e.g. bone structures appearing in CT as well as in X-ray images, or arteries and blood vessels that are visible in X-ray angiography and MR Time of Flight. Fortunately, the observation of medical images from different modalities shows particular textural and noise patterns at small scales. For instance, ultrasound images contain particular speckle patterns while relevant information is contained in low fequency edges. Hence, to recognize modalities independently of organs or anatomical structures appearing in the images, we propose to rely on local textural and noise patterns information extracted at random positions of the image. Hence, we propose the extraction of the following low-level visual features: Patch colors/intensities, Local Binary Patterns (LBP) [12], as texture operator to encode local color/intensity changes, and Histograms of Oriented Gradients (HOG) [3], to encode local appearance with local distributions of color/intensity gradient directions. These local visual features are computed on a set of patches that can be extracted densely or at particular keypoints of the image, and that may have different size. In the present work, we use patches of size 17×17 extracted at random positions. This patchsize allows us to capture small scale patterns independently of edges, corners or keypoints locations.

3.2   Extreme Random Subspace Projection Ferns

Introduced by Özuysal *et al.* in [13], Random ferns consist in an ensemble of constrained decision trees which permits to fastly partition the feature space. As shown on Fig. 2, while a tree is a set of random decision functions that split feature vectors at each node towards the left or the right branch, a fern systematically applies the same

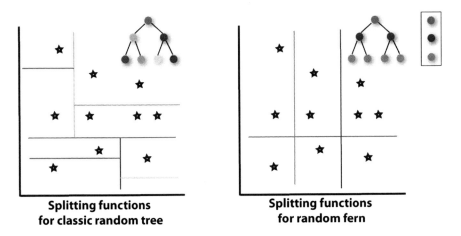

**Splitting functions
for classic random tree**

**Splitting functions
for random fern**

Fig. 2. In contrast to a tree, a fern applies only one decision function per hierachical level. It induces splitting functions which traverse the whole feature space.

decision function for each node of the current level. This means that, in contrast to random trees, the decision function is defined in the whole feature space. Results of these random tests are finally stored as binary values, leading to a more compact and simple structure for performances that are similar to those of random trees [13]. Motivated by their performance we choose to use random ferns for partitioning the visual feature space.

The ferns ensemble is built as follows: we denote by $\mathcal{F} = \left\{ \mathbf{F}^{(t)} \right\}_{t \in \mathcal{T}}$ the random ferns ensemble where $\mathcal{T} = \{1, \cdots, T\}$ are the fern indexes. Each fern $\mathbf{F}^{(t)}$ is defined as a set of L binary decision functions f_l and their associated thresholds τ_l, i.e. $\{(f_l, \tau_l)\}_{l \in \mathcal{L}}$ with $\mathcal{L} = \{1, \cdots, L\}$. The output of evaluating a pair (f_l, τ_l) on a visual feature vector $\mathbf{x}$ is binary, that is $f_l(\mathbf{x}, \tau_l) : \mathbf{x} \mapsto \{0, 1\}$. We denote the result of the evaluation b_l, where

$$\begin{cases} b_l = 0 & \text{if } f_l(\mathbf{x}) \leq \tau_l \\ b_l = 1 & \text{if } f_l(\mathbf{x}) > \tau_l \end{cases}.$$ (1)

Hence, to an input feature vector $\mathbf{x}$ corresponds a binary vector $\mathbf{b} = [b_1, \ldots, b_l, \ldots, b_L]^\top$ encoding the cell of the partition where the vector falls.

In our Extreme Random Subspace Projection Fern (ERSP) approach, we combine random dimension selection and random projections [4] at each node test. Moreover, we propose to investigate the effects of pushing the randomization one step further. Instead of searching for the best threshold according to the information gain as in Extreme Randomized Trees [5], we study the use of purely random splits. Thereby, the clustering becomes independent from the training data, providing robustness to outliers or undersampled feature spaces and which permits to generate set of thresholds that better cluster non-binary separable data once they are combined.

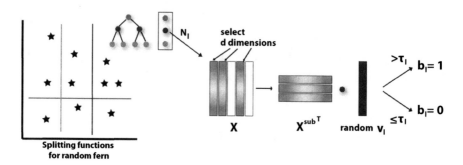

Fig. 3. ERSP algorithm: At each node, subdimensions of the original feature space are randomly selected. Then subvectors are projected using a random vector and finally, a random threshold operation is applied.

Algorithm 1. Pseudocode for Extreme Random Subspace Projection Fern

---

1: **Input**: $\{\mathbf{x}_q\}$, $(q \in \{1, \cdots, Q\})$ {input feature vectors}
2: **Output**: $\{\mathbf{b}_q\}$, {output binary vectors}
3: $\backslash\backslash$*loop over the nodes*
4: **for** each node N_l, $(l \in \mathcal{L})$ **do**
5: $\quad \backslash\backslash$*select randomly d dimensions*
6: $\quad \{\mathbf{x}_{q,l}^{\text{sub}}\} \leftarrow$ **selectRandomSubspace**$(\{\mathbf{x}_q\},d)$
7: $\quad \backslash\backslash$*generate randomly a random unit vector of dimension d*
8: $\quad \mathbf{v}_l \leftarrow$ **generateRandomProjection**
9: $\quad \backslash\backslash$*project all subvectors*
10: $\quad$ **for** each subvector $\mathbf{x}_{q,l}^{\text{sub}}$ **do**
11: $\quad\quad \mathbf{x}_{q,l}^{\text{proj}} \leftarrow \mathbf{v}_l^\top \cdot \mathbf{x}_{q,l}^{\text{sub}}$
12: $\quad$ **end for**
13: $\quad \backslash\backslash$*generate randomly a threshold in the range of the projections values*
14: $\quad \tau_l \leftarrow$ **generateRandomThreshold**$\left(\{\mathbf{x}_{q,l}^{\text{proj}}\}\right)$
15: $\quad \backslash\backslash$*perform binary test for each projection value*
16: $\quad$ **for** each projection value $\mathbf{x}_{q,l}^{\text{proj}}$ **do**
17: $\quad\quad$ **if** $\mathbf{x}_{q,l}^{\text{proj}} > \tau_l$ **then**
18: $\quad\quad\quad \mathbf{b}_q[l] = 1$
19: $\quad\quad$ **else**
20: $\quad\quad\quad \mathbf{b}_q[l] = 0$
21: $\quad\quad$ **end if**
22: $\quad$ **end for**
23: **end for**

Let us now formally describe our Extreme Random Subspace Projection Fern (ERSP) approach. We denote $\{N_l\}_{l \in \mathcal{L}}$ the nodes of a given fern. As shown on Fig. 3, at each node N_l, we first randomly select d dimensions from the visual feature space. This means at each node we only consider the set of "subvectors" $\{\mathbf{x}_{q,l}^{\text{sub}}\}$ from the subspace $\mathbb{R}^d$. Then, each subvector $\mathbf{x}_{q,l}^{\text{sub}}$ is projected to $\mathbb{R}$ using a randomly generated unit vector

$\mathbf{v}_l \in \mathbb{R}^d$: $\mathbf{x}_{q,l}^{\text{proj}} = \mathbf{v}_l^\top \cdot \mathbf{x}_{q,l}^{\text{sub}}$. The binary splitting is performed with a threshold τ_l on each projected vector $\mathbf{x}_{q,l}^{\text{proj}}$. Usually, this threshold is optimized according to the data. For instance, τ_l can be defined as the median of the projected data. In this paper, we also investigate the effects of randomizing this threshold. To summarize, the following decision function is defined $f_l(\mathbf{x}^{\text{sub}}, \mathbf{v}_l, \tau_l) : \mathbf{x}^{\text{sub}} \mapsto \{0, 1\}$:

$$f_l(\mathbf{x}^{\text{sub}}, \mathbf{v}_l, \tau_l) \doteq \max(0, \text{sign}(\mathbf{v}_l^\top \cdot \mathbf{x}^{\text{sub}} - \tau_l)).$$

The binary partition produced at node N_l is then stored in $b_l \in \{0, 1\}$. Finally, the ferns ensemble outputs a binary vector $\mathbf{b}$ encoding the index of the cell in which a feature vector falls, *i.e.* $\mathbf{b} = [b_1, \ldots, b_l, \ldots, b_L]^\top$. Note that once the training has been performed, all nodes operations are frozen. The pseudocode describing the growing of a fern is detailed in Alg. 1.

3.3 From Multiple Independent Partitions to an Implicit Dictionary

Let us denote $\{\mathcal{P}_t\}_{t \in \mathcal{T}}$ the T independent partitions of the feature space built with a ferns ensemble. Each partition $\mathcal{P}_t = \left\{\mathcal{C}_1^{(t)}, \cdots, \mathcal{C}_z^{(t)}, \cdots, \mathcal{C}_Z^{(t)}\right\}$, is defined as a set of cells of cardinality $Z = 2^L$. $\mathcal{C}_z^{(t)}$ represents the cell indexed by a unique binary vector $\mathbf{b} = [b_1, \ldots, b_l, \ldots, b_L]^\top$ resulting from the splitting operations induced by the fern $\mathbf{F}^{(t)}$. To construct our dictionary $\mathbf{D}$, all random partitions are concatenated and each of their cells are associated to a visual word of the dictionary:

$$\mathbf{D} = \{D_m\}_{m=1}^M = \left\{\mathcal{C}_1^{(1)}, \cdots, \mathcal{C}_Z^{(1)}, \cdots, \mathcal{C}_1^{(T)}, \cdots, \mathcal{C}_Z^{(T)}\right\} \qquad (2)$$

Since these visual words are not induced by an explicit global quantization of the space, but from overlapping partitions, the resulting dictionary is called "implicit". Now that the dictionary has been defined, each new feature vector $\mathbf{x}$ can be associated to a visual word as follows: First, $\mathbf{x}$ is passed through each ferns of the ERSP ensemble and corresponding output cell indexes $\{\mathcal{P}_t(\mathbf{x})\}_{t \in \mathcal{T}}$ are gathered. Then, the BoW is updated by incrementing the frequency of appearance of visual words according to these indexes. Finally, an image I is then represented by a the bag of words defined as:

$$\mathbf{h}(I) = [p(D_1|\mathcal{X}) \cdots p(D_m|\mathcal{X}) \cdots p(D_M|\mathcal{X})] \qquad (3)$$

where $\mathcal{X} = \{\mathbf{x}_q^I\}_{q=1}^Q$ is a set of features extracted from Q random patches of image I, and $p(D_m|\mathcal{X})$ the probability of a visual word D_m knowing $\mathcal{X}$. Finally, $\mathbf{h}(I)$ can be fed to a SVM classifier with a RBF kernel for modality classification.

4 Experiments and Results

In this section, we propose to recognize the modality of medical images taken from the ImageCLEF 2010 database [10]. While modality recognition was a new subtask from the medical image retrieval challenge at ImageCLEF 2010, this application brings new challenges as the database given as training set consists of classes with heterogenous content and very high variability. The same organs may appear in each modality, and different kinds of organs or anatomical structures appear within the same class. Moreover, the database contains some multi-modal images such as PET/CT, which create overlap between classes. Note that we constrain here the problem to the most interesting and challenging modalities from the database:

- **CT**: Computerized tomography (314 images).
- **MR**: Magnetic resonance imaging (299 images).
- **PET**: Positron emission tomography including PET/CT (285 images).
- **US**: ultrasound including (color) doppler (307 images).
- **XR**: x-ray including x-ray angiography (296 images).

This leads to a total of 1501 images. **GX** (Graphics,drawings,...) and **PX** (Photographs,...) classes have been discarded for our study.

To build the dictionaries, 5000 random patches are extracted for each class which make a total of 25000 visual features. During the test phase, 1000 random features are computed to construct the BoW of an image. For each test, classes are rebalanced by using random subsampling. A grid-search is performed to find the best hyperparameters for the SVM classifier. Note that all experiments are performed with MATLAB and we use the fast Kmeans++ implementation proposed by Arthur *et al.*[1] to perform K-means clustering with a clever initialization.

Tab. 1 compares the classification results and Tab. 2 the times needed to cluster the data points to create the dictionary. Our approach performs slightly better than

Table 1. Classification results of our approach against K-means for different numbers of ferns and nodes. We investigate the effects of randomizing the choice of the threshold (results in parenthesis) in our approach.

CLASSIFICATION RESULTS

K-means				
Nb of clusters	1000	2000	5000	10000
F-measure	75.1%	**75.2%**	75%	74.2%

Our approach				
ferns/nodes/clusters	8/8/2048	8/10/2560	10/8/8192	10/10/10240
F-measure	76.8(74.9)%	75.8(75.1)%	76.8(75.9)%	**76.9(76)**%

Table 2. Clustering time for our approach against K-means for different numbers of ferns and nodes

CLUSTERING TIME

K-means				
Nb of clusters	1000	2000	5000	10000
Time in *hours*	2.3 h	**3.5 h**	6.6 h	11.3 h

Our approach				
ferns/nodes/clusters	8/8/2048	8/10/2560	10/8/8192	10/10/10240
Time in *seconds*	1.08 s	1.16 s	1.28 s	**1.41 s**

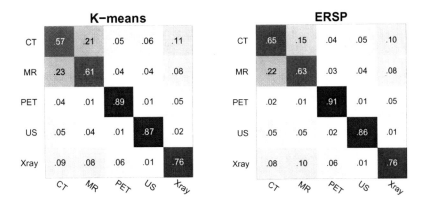

Fig. 4. Confusion matrices for K-means (left) compared to our approach (right)

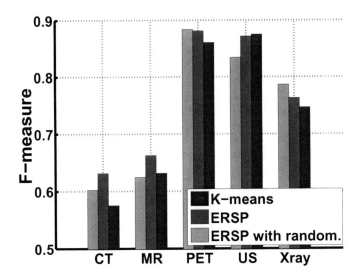

Fig. 5. comparative results of the overall classification accuracy

K-means even if thresholds are randomly chosen. Moreover, while K-means needs several hours to perform one clustering run, our method clusters the data in less than two seconds. Fig. 4 compares the confusion matrices our approach against K-means, and Fig. 5 presents the classification results for each class. For both approaches, most of the confusion occurs between the CT and MR, and between the two and some of the X-ray images. This is expected as CT and MR may sometimes be difficult to discriminate using only local information. Indeed, they both contain patterns showing high variability according to the chosen feature representation. Moreover, images may suffer from artifacts due to the imaging system itself or to jpeg compression. Such artifacts may alter intensity patterns and distributions or worst, create artificial structures that look very

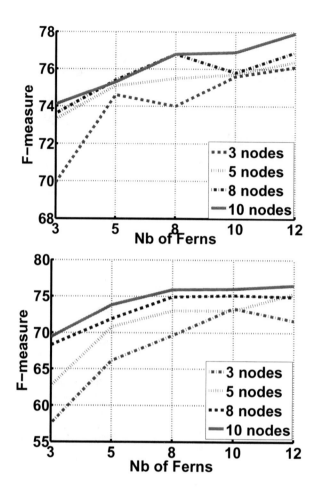

Fig. 6. Classification results for ERSP approach without (top) and with (bottom) threshold randomization according to the number of ferns and to the number of nodes

similar in CT and MR. Confusions between CT and X-ray can be explained from the fact that they are based on the same physical phenomenon. Concerning MR and X-ray, confusions occur for instance in the case of cropped images of the knee which, except from the cartilage, are very dark. On the other hand, PET and US images show very discriminant patterns that can be very well separated. In the case of PET-CT images, since colors are used to represent PET signals, they can be be also well recognized. In Fig. 6, we compare the influence of increasing the number of ferns on the f-measure, and this for both ERSP methods with and without threshold randomization. These figures suggest that with extreme randomization: (1) more ferns are needed to achieve comparable performance, and (2), the performance converges towards a limit while in the other case, we can expect further increase in the f-measure.

5   Discussion and Conclusion

In this paper, our contribution is an approach to construct implicit dictionaries for modality recognition using extreme randomization. The backbone of our method is a clustering algorithm based on random ferns we call Extreme Random Subspace Projection (ERSP) ferns. Our approach is very fast, it provides independence from the available training set through extreme randomization, it is not highly dependent on the initialization, and it does not require the a-priori knowledge of the number of clusters. Experiments conducted on medical images from ImageCLEF 2010 database show that our approach is a fast alternative to K-means clustering for building efficiently dictionaries for multi-class classification.

References

[1] Arthur, D., Vassilvitskii, S.: K-means++: The advantages of careful seeding. In: SODA (2007)
[2] Cortes, C., Vapnik, V.: Support vector networks. Mach. Learn. 20, 273–297 (1995)
[3] Dalal, N., Triggs, B.: Histograms of oriented gradients for human detection. In: CVPR (2005)
[4] Freund, Y., Dasgupta, S., Kabra, M., Verma, N.: Learning the structure of manifolds using random projections. In: NIPS (2007)
[5] Geurts, P., Ernst, D., Wehenkel, L.: Extremely randomized trees. Mach. Learn. (2006)
[6] Jurie, F., Triggs, B.: Creating efficient codebooks for visual recognition. In: ICCV (2005)
[7] Kalpathy-Cramer, J., Hersh, W.: Multimodal medical image retrieval: image categorization to improve search precision. In: Int. Conf. on Multimedia Information Retrieval (2010)
[8] Mairal, J., Bach, F., Ponce, J., Sapiro, G., Zisserman, A.: Supervised dictionary learning. In: NIPS (2008)
[9] Moosmann, F., Triggs, B., Jurie, F.: Fast discriminative visual codebooks using randomized clustering forests. In: NIPS (2006)
[10] Müller, H., Kalpathy-Cramer, J., Eggel, I., Bedrick, S., Kahn Jr., C.E., Hersh, W.: Overview of the clef 2010 medical image retrieval track. In: Image CLEF (2010)
[11] Nistér, D., Stewénius, H.: Scalable recognition with a vocabulary tree. In: CVPR (2006)
[12] Ojala, T., Pietikinen, M., Harwood, D.: A comparative study of texture measures with classification based on feature distributions. In: Pattern Recognition (1996)
[13] Özuysal, M., Calonder, M., Lepetit, V., Fua, P.: Fast keypoint recognition using random ferns. In: PAMI (2010)
[14] Perbet, F., Stenger, B., Maki, A.: Random forest clustering and application to video segmentation. In: BMVC (2009)
[15] Perronnin, F., Dance, C.R., Csurka, G., Bressan, M.: Adapted Vocabularies for Generic Visual Categorization. In: Leonardis, A., Bischof, H., Pinz, A. (eds.) ECCV 2006. LNCS, vol. 3954, pp. 464–475. Springer, Heidelberg (2006)
[16] Shotton, J., Johnson, M., Cipolla, R.: Semantic texton forests for image categorization and segmentation. In: CVPR (2008)
[17] Sivic, J., Zisserman, A.: Video google: A text retrieval approach to object matching in videos. In: ICCV (2003)
[18] Winn, J., Criminisi, A., Minka, T.: Object categorization by learned universal visual dictionary. In: ICCV (2005)
[19] Yang, L., Jin, R., Sukthankar, R., Jurie, F.: Unifying discriminative visual codebook generation with classifier training for object category recognition. In: CVPR (2008)

Superpixel-Based Interest Points for Effective Bags of Visual Words Medical Image Retrieval

Sebastian Haas[1,*], René Donner[1], Andreas Burner[1], Markus Holzer[1], and Georg Langs[1,2]

[1] Computational Image Analysis and Radiology Lab, Department of Radiology, Medical University of Vienna, Austria
sebastian.haas@meduniwien.ac.at
[2] Computer Science and Artificial Intelligence Laboratory, Massachusetts Institute of Technology, Cambridge, MA, USA

Abstract. The present work introduces a 2D medical image retrieval system which employs interest points derived from superpixels in a bags of visual words (BVW) framework. BVWs rely on stable interest points so that the local descriptors can be clustered into representative, discriminative prototypes (the visual words). We show that using the centers of mass of superpixels as interest points yields higher retrieval accuracy when compared to using Difference of Gaussians (DoG) or a dense grid of interest points. Evaluation is performed on two data sets. The Image-CLEF 2009 data set of 14.400 radiographs is used in a categorization setting and the results compare favorable to more specialized methods. The second set contains 13 thorax CTs and is used in a hybrid 2D/3D localization task, localizing the axial position of the lung through the retrieval of representative 2D slices.

Keywords: medical image retrieval, bags of visual words, superpixel, interest points.

1 Introduction

In clinical practice, radiologists rely mainly on their experience to form diagnoses. Typically the vast amount of imaging data in hospitals is not exploited due to the lack of easy to use and effective image retrieval systems. Ideally, such systems should present the radiologists with comparable cases to the one in question. In this work, we present a 2D medical image retrieval system which advances on the state of the art by incorporating a novel kind of interest points which is shown to yield very stable results on medical data.

The bags of visual words approach, for which [13] provides a good general overview, has shown promising results in medical image retrieval tasks [6,3]. The main trend is the shift from purely Difference of Gaussians based interest

---

* The research leading to these results has received funding from the Austrian Sciences Fund (P 22578-B19, PULMARCH) and the EU-funded KHRESMOI project (FP7-ICT-2009-5/257528).

H. Müller et al. (Eds.): MCBR-CDS 2011, LNCS 7075, pp. 58–68, 2012.

points to dense grids of interest points - thus effectively eliminating the influence of any interest point detector, altogether [2,1]. The stability of the interest point detector is crucial for obtaining discriminative descriptor prototypes in the clustering stage of the bags of visual words approach. While the most popular descriptors (e.g., SIFT [8], or derivatives such as SURF [4]) are insensitive to small perturbations in the interest points, fine-grained, repetitive structures like those encountered in microscopic images [2] or lung CTs [6] prove difficult for repeatable interest point identification.

1.1 Related Work

The performance of several image retrieval and annotation frameworks is compared during the ImageCLEF medical tracks. Several groups have proposed methods based on BVWs in this context. *TAUbiomed* based a BVW dictionary on patches of 9x9 pixels in the images, and performed classification by a support vector machine (SVM) using the word histograms [3]. Dimitrovski et al. proposed a hierarchical multi-label classification, using edge histogram descriptors as global and SIFT descriptors and Local Binary Patterns (LBP) as local features [5]. Unay et al. used LBP as features together with hierarchical SVMs [10]. However, none of these groups have combined the bags of visual words approach with superpixels as interest points for SIFT descriptors.

In the following Sections we outline the concept of bags of visual words and describe superpixel based interest points. In Section 3 the evaluation setup is detailed with the results presented in Section 4. We end the paper by providing a conclusion and an outlook.

2 Methods

We set up a flexible bags of visual words framework and replace selected parts of the BVW workflow for comparison and evaluation. In this specific case we test different interest point detectors. Section 2.1 describes the concept of the BVW framework. The three interest point detectors, we compare, are then described in Section 2.2.

2.1 Bags of Visual Words

BVW uses a set of training images to generate a so-called visual vocabulary derived from local image descriptors. The vocabulary is computed through clustering of all descriptors calculated from the training images. During retrieval both the training images and the query images can then be represented by a histogram of the visual words, and the distances between images correspond to distances between the histograms.

More formally, for each image $\mathbf{I}_i$ in the set of training images $\mathcal{I} = \mathbf{I}_1, \ldots, \mathbf{I}_N$ K_i interest points $\mathbf{i}_j$ are computed. Each is comprised of coordinates x_j and y_j, a scale s_j and an orientation θ_j,

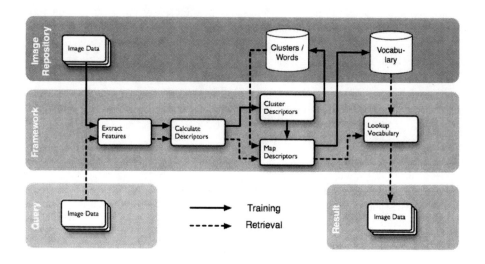

Fig. 1. Schematic overview of the Bag of Visual Words framework

$$\mathbf{i}_j = (x_j, y_j, s_j, \theta_j)^T. \tag{1}$$

Corresponding with each $\mathbf{i}_j$ a local descriptor $\mathbf{d}_j$ is extracted, e. g. the $d = 128$ dimensional SIFT descriptor [8]. Assembling all descriptors of all images into the $d \times D$ descriptor space $\mathbf{D}$ allows to perform clustering. Typically k-means clustering is used. The resulting k d-dimensional cluster centers $\mathbf{c}_l$ represent the visual words.

To obtain the histogram $\mathbf{h}_i$ for image $\mathbf{I}_i$ we perform vector quantization on all its descriptors $\mathbf{d}_j$ onto the cluster centers $\mathbf{c}_l$: Each $\mathbf{d}_j$ is assigned the label l of the closest cluster, using Euclidean distance in the descriptor space. Forming a histogram over all the labels encountered within an image yields the k-bin histogram $\mathbf{h}_i$, which is normalized to sum to 1.

Retrieval The retrieval step is similar to the processing of one training image. The interest points and corresponding descriptors are extracted for the query image $\mathbf{I}_q$ and mapped to the visual words. The histogram $\mathbf{h}_q$ based on the cluster mapping is used in a nearest neighbor query using the $\chi^2 - distance$ against the database of previous stored histograms.

Figure 1 shows the training and retrieval pipeline used in the BVW framework.

2.2 Interest Points: DoG, DENSE, Superpixel Points

We evaluate the influence of varying the type of interest points on the quality of the BVW retrieval result. We compare SIFT interest points (DoG keypoints), DENSE points (regular grid) and superpixel points.

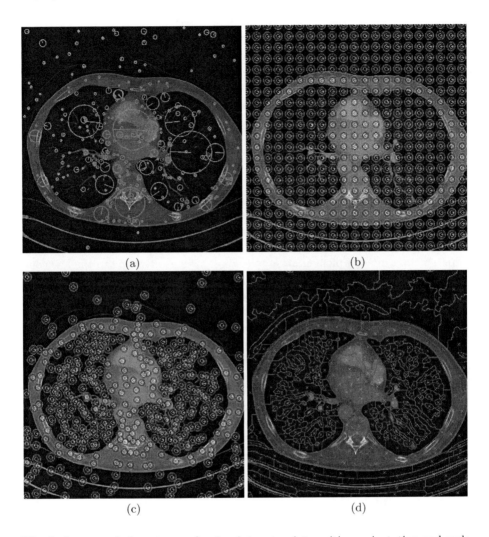

Fig. 2. An example lung image showing interest point position, orientation and scale for a) SIFT b) DENSE c) superpixel d) shows the the superpixel regions as well as the centers of mass used as interest points

SIFT: The Scale-Invariant Features introduced by Lowe are widely used in the field of image processing. Being invariant against scaling, rotation or translation of the image helps finding similar image regions in manipulated or even different images. We use the standard implementation provided by [11] to perform our experiments. An example for SIFT interest points showing their location, scale and orientation can be found in Figure 2 (a).

DENSE: Dense SIFT as used in [2] is based on a regular grid over the image. The interest points used for the computation of the SIFT descriptors are generated based on the size of the image. The content of the image does not affect the DENSE interest points. The interest points are located on a regular grid relative to the image coordinate system. Constant values are used for the orientation and scale needed for the computation of SIFT descriptors. In our setup we assign each point an orientation of 0° and scales of 4, 6, 8 and 10 pixels leading to four interest points per location. In contrast to Andreé [2], our interest points at not overlapping. We use a rather wide grid spacing of 25 pixels to reduce the amount of interest points to a similar value like SIFT or superpixels compute. The dense grid of interest points is shown in Figure 2 (b).

Superpixel points: Superpixels represent an over-segmentation of the underlying image into partitions which exhibit maximal internal homogeneity and maximal inter-super pixel heterogeneity, while respecting image boundaries. Several different approaches for conforming with this loose definition have been proposed. They vary in terms of smoothness, regularity and computational cost. In this paper, a superpixel implementation following [12] was used due to the low computational cost and the low number of resulting superpixels in background areas. The chosen method employs the monogenic signal to compute the local phase. This measure is invariant to changes in contrast and brightness. Local phase means that each pixel contains directional information to the strongest edge within a radius r corresponding to the selected scale of the monogenic signal. This property makes the monogenic signal very well suited for medical image data, with their often subtle edge gradient which defy conventional edge detectors. A watershed transform on the monogenic signal yields the regions which are used as superpixels in our approach.

To obtain interest points from these regions the center of mass of each region is computed. Again, the orientation is set to 0° and four interest points with a scales of 4, 6, 8 and 10 pixels are created per center.

Figure 2 (c) shows the superpixel interest points for an example lung slice. In (d) the borders and centers of mass are shown for all superpixel regions.

2.3   Distribution of Interest Points in the Image

Figure 3 shows an example slice of a lung CT (a) and the distribution of SIFT (b) and superpixel (c) interest points on this image. The distribution of the interest points is shown as a heat map image with 25 by 25 bins. Each bin represents an area of approximately 20 by 20 pixels of the original image. The heat map for the DENSE interest points was left out because of their uniform distribution.

The superpixel heat map shows a lower density in the upper part of the image. This area shows mainly the air above the thorax. The superpixel algorithm generates in this area fewer, but bigger superpixels. Compared to SIFT the superpixel algorithm generates in the area of the lung more uniform distributed interest points.

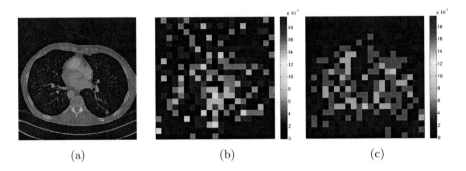

(a)   (b)   (c)

Fig. 3. Based on the image shown in a) the distribution of the SIFT and superpixel interest points are shown in b) and c), respectively. The number of interest points per type are normalized to obtain comparable heat maps. Note the more even and more concentrated distribution in the area of the medical object of interest.

3   Experiments

In this section we describe two independent experiments for testing the three types of interest points.

The first experiment uses the bags of visual words framework for an annotation task. This task is based on the ImageCLEF 2009 Medical Image Annotation Task [9]. The used image repository[1] contains 14.410 x-ray images split into a training set (12.677 images) and a test set (1.733 images). All images are fully annotated with the IRMA code [7].

The IRMA code for each test set image is derived from the retrieval result image with the highest similarity. We assign each query image the IRMA code corresponds to the closest result image in the training data. The classified images were evaluated with the scheme[2] used by the ImageCLEF 2009 Medical Image Annotation Task. Although this scheme allows wildcards within the IRMA code, we don't use them. Our experiment only takes the classification set *2008* into account, which is the most complete classification and based on the IRMA code. The other three sets *2005, 2006* (both are sets of simple raw numbers) and *2007* (IRMA code) are not used.

The second experiment uses the BVW framework for locating CT slices of lung CTs based on the position of the retrieved images. Histograms of visual words in combination with dense SURF descriptors have been successfully used for estimating body portions using CT volumes[6]. For this run we use 13 lung CTs, each CT containing between 132 and 530 slices with a resolution of 512×512 pixel. The slices of each CT containing parts of the lung are registered on a scale from 0.0 to 1.0 where the proximal occurrence of the lung in the volume has the value 0.0 and the distal occurrence has the value 1.0. These two key-images

---

[1] http://www.irma-project.org/datasets.php?SELECTED=00009#00009.dataset
[2] http://www.idiap.ch/clef2009/evaluation_tools/error_evaluation.pdf

are selected manually. All slices in between get equally distributed coordinates assigned. The slices cranial of the first lung-slice get negative values and the slices caudad of the last lung-slice get values greater than 1.0. Figure 4 shows an ordered subset of images taken from one CT with their assigned position value. The first and last occurrence of the lung are marked. The 13 lung CTs are processed with the BVW framework using a leave-one-out cross-validation. From every CT 100 images with equally spaced positions between 0.0 and 1.0 are selected as query images. Based on this query images we test the location accuracy for the slices of the whole lung CT. The position is determined by the median position value of the 10 most similar retrieved images per query.

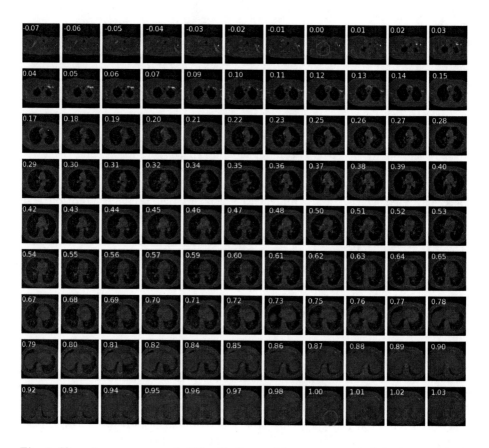

Fig. 4. Slices from one example CT with first and last occurrence of the lung marked. The numbers indicate the registered positions of the slices.

Both experiments are set up for SIFT, DENSE and superpixel interest points. The BOW dictionary is build from 1000 images leading to approximately $1.6*10^6$ to $2.0*10^6$ interest points and descriptors. These descriptors are clustered to 1000 words using k-means.

4   Results

The error scores of the first experiments are calculated by using the scoring schema defined for the ImageCLEF 2009 Medical Image Annotation Task. The superpixel approach clearly outperforms the approaches with DENSE and SIFT interest points with a score of 249.34 against 442.51 and 562.75, respectively. Table 1 shows the best run of each competitor of the ImageCLEF in addition to the three approaches evaluated in this paper.

Although the superpixel approach shows a good performance, we expect a better values by improving the selection of IRMA code by e. g. taking more than the first retrieved image into account.

Table 1. This table shows the error score of the three approaches described in this paper (bold) and the best result of each group from the ImageCLEF 2009 Medical Image Annotation Task [9]. The error score is based on the evaluation scheme for the IRMA code classification set *2008*. A low score indicates a higher accuracy.

Approach	error score
TAUbiomed	169.50
Idiap	178.93
FEITIJS	242.46
Superpixels	**249.34**
VPA SabanciUniv	261.16
MedGIFT	317.53
IRMA	359.29
DENSE	**442.51**
SIFT	**562.75**
DEU	642.50

For the second experiment, for each leave-one-out run, we retrieve the 10 most similar images for each of the the 100 images from test CT. The position is determined by the median position value of the ten best retrieved images per query. Knowing the position of each query image in the ground truth, we are able to aggregate the computed image-positions over all 13 CTs. The 13 leave one out runs are aggregated by computing the mean position for each of the 100 test slices. Figure 5 shows plots of the mean value and standard deviation of the aggregated data for the SIFT, DENSE and superpixel interest points. The x-axis of the plots defines the image position in the ground truth whilst the mean value and the standard deviation is plotted against the y-axis. The baseline shown as the 45 ° line indicates the optimal result. The plots show that the mean values of the superpixel image-positions are closer to the baseline than in the results of the other two runs.

The overall performance of the three interest point types is summed up on Table 2. Superpixels with a sum-of-squared-erros of 0.0554 perform better than SIFT and DENSE interest points with 0.1416 and 0.1453 respectively.

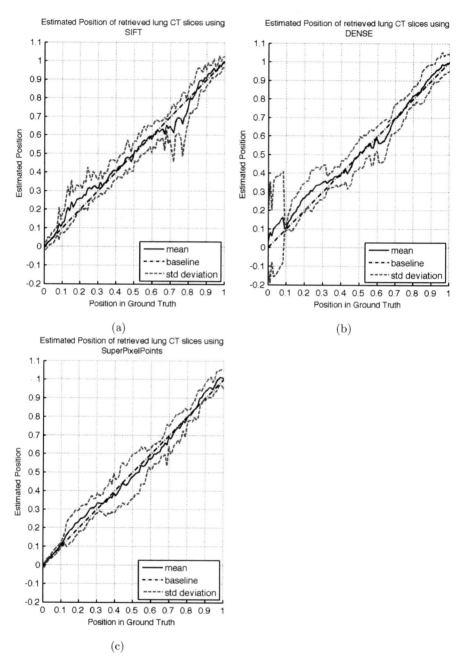

Fig. 5. Results of location retrieval using (a) SIFT, (b) DENSE and (c) superpixel interest points. The known position of the 100 query images are plotted on the x-axis. The y-axis shows the computed image position. The plots show the optimal result (baseline), the mean position and standard deviation for each query position aggregated over the 13 CTs.

Table 2. Aggregated deviation of our test results shown in Figure 5. Mean, median and sum of squared errors (SSE) for the distance between our results and the baseline of the three lung CT position estimation runs.

Interest points	mean	median	SSE
SIFT	0.0278	**0.0186**	0.1416
DENSE	0.0295	0.0233	0.1453
Superpixel	**0.0199**	0.0234	**0.0554**

5 Conclusion

In this paper we have introduced and evaluated a medical image retrieval and localization system which extends the Bags of Visual Words approach to using interest points derived from superpixels. Superpixels provide a stable oversegmentation of image structures, and their centers of mass thus yield stable interest points. This is particularly relevant for medical imaging data, where both salient points, and homogeneous structures occur at unique locations across a population. This in turn allows the clustering stage in BVWs to find discriminative prototypes which improve the retrieval accuracy of the proposed system.

For both tasks, the classical image category classification and the 2D/3D localization task, superpixel interest points yielded better results, while being straightforward and fast to compute. Future work will evaluate different types of superpixels for their stability and extend the proposed retrieval system to 3D data.

References

1. André, B., Vercauteren, T., Perchant, A., Wallace, M.B., Buchner, A.M., Ayache, N.: Endomicroscopic image retrieval and classification using invariant visual features. In: Proceedings of the Sixth IEEE International Symposium on Biomedical Imaging 2009 (ISBI 2009), pp. 346–349. IEEE, Boston (2009)
2. André, B., Vercauteren, T., Wallace, M.B., Buchner, A.M., Ayache, N.: Endomicroscopic video retrieval using mosaicing and visual words. In: Proceedings of the Seventh IEEE International Symposium on Biomedical Imaging 2010 (ISBI 2010), pp. 1419–1422. IEEE (2010)
3. Avni, U., Goldberger, J., Greenspan, H.: Addressing the ImageCLEF 2009 challenge using a patch-based visual words representation. In: Working Notes for the CLEF 2009 Workshop. The Cross-Language Evaluation Forum (CLEF), Corfu, Greece (2009)
4. Bay, H., Tuytelaars, T., Van Gool, L.: SURF: Speeded Up Robust Features. In: Leonardis, A., Bischof, H., Pinz, A. (eds.) ECCV 2006. LNCS, vol. 3951, pp. 404–417. Springer, Heidelberg (2006)
5. Dimitrovski, I., Kocev, D., Loskovska, S., Džeroski, S.: ImageCLEF 2009 Medical Image Annotation Task: PCTs for Hierarchical Multi-Label Classification. In: Peters, C., Caputo, B., Gonzalo, J., Jones, G.J.F., Kalpathy-Cramer, J., Müller, H., Tsikrika, T. (eds.) CLEF 2009. LNCS, vol. 6242, pp. 231–238. Springer, Heidelberg (2010)

6. Feulner, J., Zhou, S.K., Seifert, S., Cavallaro, A., Hornegger, J., Comaniciu, D.: Estimating the body portion of CT volumes by matching histograms of visual words. In: Medical Imaging 2009: Image Processing (Proceedings Volume), vol. 7259, p. 72591. SPIE (2009)

7. Lehmann, T.M., Schubert, H., Keysers, D., Kohnen, M., Wein, B.B.: The IRMA code for unique classification of medical images. In: Medical Imaging 2003: PACS and Integrated Medical Information Systems: Design and Evaluation (Proceedings Volume), vol. 5033, pp. 440–451. SPIE (2003)

8. Lowe, D.G.: Distinctive image features from scale-invariant keypoints. International Journal of Computer Vision 60, 91–110 (2004)

9. Tommasi, T., Caputo, B., Welter, P., Güld, M.O., Deserno, T.M.: Overview of the CLEF 2009 Medical Image Annotation Track. In: Peters, C., Caputo, B., Gonzalo, J., Jones, G.J.F., Kalpathy-Cramer, J., Müller, H., Tsikrika, T. (eds.) CLEF 2009. LNCS, vol. 6242, pp. 85–93. Springer, Heidelberg (2010)

10. Ünay, D., Soldea, O., Akyüz, S., Çetin, M., Erçil, A.: Medical image retrieval and automatic annotation: VPA-SABANCI at ImageCLEF 2009. In: Working Notes for the CLEF 2009 Workshop. The Cross-Language Evaluation Forum (CLEF), Corfu, Greece (2009)

11. Vedaldi, A., Fulkerson, B.: Vlfeat: an open and portable library of computer vision algorithms. In: Proceedings of the International Conference on Multimedia, MM 2010, pp. 1469–1472. ACM, New York (2010)

12. Wildenauer, H., Mičušík, B., Vincze, M.: Efficient Texture Representation using Multi-Scale Regions. In: Yagi, Y., Kang, S.B., Kweon, I.S., Zha, H. (eds.) ACCV 2007, Part I. LNCS, vol. 4843, pp. 65–74. Springer, Heidelberg (2007)

13. Zhang, J., Marszalek, M., Lazebnik, S., Schmid, C.: Local features and kernels for classification of texture and object categories: A comprehensive study. In: Conference on Computer Vision and Pattern Recognition Workshop, p. 13 (June 2006)

Using Multiscale Visual Words for Lung Texture Classification and Retrieval

Antonio Foncubierta-Rodríguez[1,*], Adrien Depeursinge[1,2],
and Henning Müller[1,2]

[1] University of Applied Sciences Western Switzerland (HES–SO), Sierre, Switzerland
antonio.foncubierta@hevs.ch
[2] University and University Hospitals of Geneva (HUG), Switzerland

Abstract. Interstitial lung diseases (ILDs) are regrouping over 150 heterogeneous disorders of the lung parenchyma. High–Resolution Computed Tomography (HRCT) plays an important role in diagnosis, as standard chest x–rays are often non–specific for ILDs. Assessment of ILDs is considerd hard for clinicians because the diseases are rare, patterns often look visually similar and various clinical data need to be integrated. An image retrieval system to support interpretation of HRCT images by retrieving similar images is presented in this paper. The system uses a wavelet transform based on Difference of Gaussians (DoG) in order to extract texture descriptors from a set of 90 image series containing 1679 manually annotated regions corresponding to various ILDs. Visual words are used for feature aggregation and to describe tissue patterns. The optimal scale–progression scheme, number of visual words, as well as distance measure for clustering to generate visual words are investigated. A sufficiently high number of visual words is required to accurately describe patterns with high intra–class variations such as healthy tissue. Scale progression has less influence and the Euclidean distance performs better than other distances. The results show that the system is able to learn the wide intra–class variations of healthy tissue and the characteristics of abnormal lung tissue to provide reliable assistance to clinicians.

Keywords: Chest, CT, Computer–aided diagnosis.

1 Introduction

The use of images in medicine to support diagnosis has followed an exponential growth over the past 20 years [16]. New imaging techniques provide useful information to radiologists and other clinicians leading to an accurate diagnosis without a need for invasive techniques such as biopsies [20]. As a consequence, radiologists face changing problems: several studies have shown a high inter– and intra–observer variability in image–based diagnosis but a reduced variability for experienced observers [2,11]. Quick access to similar cases of the past with accurate diagnosis and further case information appears as useful to less

---

* Corresponding author.

H. Müller et al. (Eds.): MCBR-CDS 2011, LNCS 7075, pp. 69–79, 2012.

experienced radiologists in order to make decisions consistent [1]. This leads to the second problem: the ever–increasing amount of images available makes it difficult to find relevant similar cases that can improve diagnosis. Without using efficient computerized image search, this task becomes time–consuming and observer–dependent such as searching cases in text books. Content–based image retrieval (CBIR) systems provide an objective, consistent similarity measure for visual retrieval and provide results to radiologists quickly, often allowing for interactive retrieval. CBIR provides radiologists with useful, related information of similar cases helping them make decisions based on the synthesis of similar cases and their own knowledge. Many CBIR systems focus on medical data to make relevant information accessible for clinicians [15,1,14].

Interstitial lung diseases (ILDs) group over 150 lung disorders characterized by gradual alteration of lung parenchyma leading to breathing dysfunction. The diagnosis is based on a combination of patient history, physical examination, pulmonary function testing and other analyses such as blood tests. When the synthesis of this suggests an ILD, high–resolution computer tomography (HRCT) plays an important role in establishing a differential diagnosis. HRCT gives radiologists the opportunity to visualize and identify the abnormalities occurring in the lung parenchyma based on textures of the lung tissue [18,9]. Pattern recognition and classification of abnormalities are essential skills for identifying ILDs [2]. In this paper, lung image retrieval is presented that is capable of enabling quick access to similar cases based on the texture of lung regions. The technique presented in this article uses distributions of multi–scale visual words to characterize the specific texture signatures of six lung tissue types commonly found in ILDs. It differs from previous efforts with similar techniques [10] in the size of the database, the classification of a higher number of tissue types and the inclusion of multi–scale information for the visual words. The optimal number of visual words is investigated. The size of the visual vocabulary is known to be a trade–off between discriminativity and generalizability [19]. A similar challenge occurs for the scale progression, which may have an impact on the texture description [6]. The main contribution of this paper is a systematic analysis of visual vocabularies of varying size within a wavelet–based framework. The optimal number of visual words, scale progression as well as distance measures used for the clustering to generate the vocabulary are investigated. Although the application domain of this paper are ILDs, it can be extended to other image types with texture-based information, such as liver tissue in CT images.

2   Methods

2.1   Database

This work is based on a multimedia database of ILD cases created at the Geneva University Hospitals within the context of the Talisman [1] project [7]. A set of

---

[1] Talisman: Texture Analysis of Lung ImageS for Medical diagnostic AssistaNce, www.sim.hcuge.ch/medgift/01_Talisman_EN.htm as of 15 December 2010.

90 thoracic HRCT scans of 85 patients (slice thickness 1mm, inter–slice distance 10mm) having 1679 annotated image regions is used. This represents the most frequent disease patterns. Healthy and five pathological lung tissue types commonly used to characterize frequent ILDs in HRCT are used as texture classes (i.e. consolidation, emphysema, fibrosis, ground glass, healthy and micronodules, see Figure 2).

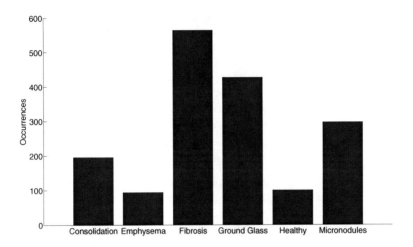

Fig. 1. Uneven distribution of disease patterns in the database

The image database contains hand–drawn regions annotated in a collaborative fashion by two radiologists with 15 and 20 years of experience. These annotations are used as ground truth for performance evaluation. The diagnosis of each case was confirmed either by pathology (biopsy, bronchoalveolar washing) or by a laboratory/specific test. Among the annotated regions in the database, six tissue patterns are sufficiently frequently found for an automated analysis but the frequencies are not evenly distributed. Healthy tissue was obtained for most patients but as many are elder persons there often is very little healthy tissue. Figure 2 shows examples of patterns observed in the lung tissue and their distribution in the dataset. (see Figure 1).

2.2   Techniques Applied

The proposed CBIR system is based on two main ideas: a wavelet transform is used to provide multi–scale representations of the lung texture; visual words are computed from the wavelet coefficients to reduce dimensionality and to build texture descriptors based on patterns actually occurring in the data.

The wavelet–based analysis is performed with focus on the scale parameters, since previous efforts have shown the influence of scale progression in image retrieval [6]. Systematic scale progression variations allow obtaining reliable

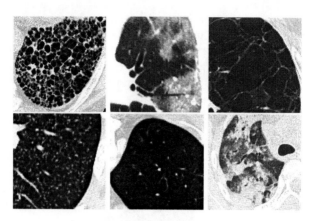

Fig. 2. Examples for texture in lung tissue showing (from top left to bottom right): fibrosis, ground glass, emphysema, micronodules, healthy and consolidation tissue

information on the trade–off between information redundancy and retrieval performance. For the wavelet analysis, mother and father wavelets are Gaussian. This multi–scale analysis yields high–dimensional feature spaces, with 5 to 14 subbands based on the scale progression chosen. In order to reduce dimensionality and to describe the information contained in a region of interest rather than the texture information of a pixel neighborhood, clustering of the wavelet coefficients is performed. Each cluster centroid then corresponds to a so–called visual word. To characterize local texture properties, each pixel of the annotated region is described by the closest centroid. This allows to characterize an entire region by a histogram of the visual words of its pixels.

Wavelet Analysis. Wavelet theory consists of a framework allowing for multi–resolution and multi–scale analysis of images by convolution with scaled and translated versions of a bandpass function, the mother wavelet ψ (see Equation 1). This scaling in the spatial domain corresponds to another scaling in the frequency domain, as shown in Equation 2, where $\Psi(\omega)$ denotes the Fourier transform of $\psi(t)$.

$$\psi_{s,\tau}(t) = \frac{1}{\sqrt{s}} \psi\left(\frac{t - \tau}{s}\right). \tag{1}$$

$$\Psi_{s,\tau}(\omega) = \frac{1}{\sqrt{s}} |s| \, \Psi(s\omega) \, e^{-j\omega\tau}. \tag{2}$$

From Equation 2 it can be seen that for each value of s, a bandpass filter of bandwidth B/s can be obtained, being B the bandwidth of the mother wavelet.

A common formulation for the scale definition is based on the minimum number of scales that guarantee an invertible transform. Scales s change according to a dyadic scheme, where $s = 2^j$ defines the scale at iteration j. This leads to a division of the frequency domain in octaves. However, texture analysis may require intermediate scales as they can improve retrieval performance [6]. To define intermediate scales, a varying number of voices v per octave can be used to define the scale parameter $s = 2^{j/v}$.

To reduce the number of scales needed for analyzing the whole frequency spectrum, the wavelet theory defines the concept of father wavelet, also known as scaling function, which consists of the lowpass filter corresponding to a certain scale.

In this text, several scale progressions were tested by using different values for the number of voices v. Only one subband of the scaling function per progression scheme is used, and both the mother and father wavelets are derived from Gaussians. The scaled bandpass filters (i.e., wavelets) are expected to have their maximum response for different lung texture types. A fully isotropic scaling function is used to extract the low pass information of the texture. For this purpose, a 2D Gaussian filter is chosen. Mexican Hat wavelets are used to provide band–limited isotropic texture analysis. This work uses a precise approximation of the Mexican Hat wavelet that also satisfies the wavelet admissibility condition while being easier to compute, the Difference of Gaussians (DoG).

Visual Words. The term *texture* often has a fuzzy definition and refers to the characteristics of the pixel values within a certain region and their relationships. Since the wavelet transform can describe the transient of the values in the voxel surroundings, a way of aggregating this information for a region of interest is needed. Visual words [17] have been widely used in image retrieval and classification for describing images (or regions of interest) similarly to the bag–of–words approach used for text retrieval. For each voxel, this technique maps a set of continuous low–level features, e.g. gray values or wavelet coefficients, into a compact discrete representation consisting in visual words. Every voxel is now described by a single word instead of a set of features. The aggregation of visual words for describing a region in an image is carried out by calculating the histogram of the words that appear within it, providing high–level features. The Multi–Scale visual–words (MSVW) aim at characterizing the organization of the voxel values within a region. Whereas in text retrieval a document is described as the histogram of word occurrences, a given vocabulary in CBIR systems has to be created automatically from image data based on clustering of visual properties.

Several techniques have been proposed in the literature to quantize descriptors into visual words [3,17]. There are various clustering approaches depending on the desired characteristics for the clusters. In this paper, the well–known *k–means* clustering is used. K–means aims at finding clusters iteratively, assuming a (hyper–) spherical cluster model and that all clusters are approximately of the same size. For this reason the cluster assignment is done by selecting the nearest cluster centroid in terms of Euclidean distance. Adding a feature with a large variability such as the original gray level value can make data too sparse and therefore difficult to cluster. Other distance measures are explored for removing inter–feature correlation, such as the Mahalanobis distance [13] or by normalizing the maximum value of features prior to clustering.

Table 1. Features for every scale progression. ϕ corresponds to the scaling function values, whereas ψ_s corresponds to the wavelet values for scale s

	1 voice	1.5 voices	2 voices	2.5 voices	3 voices	3.5 voices	4 voices
Feature 1	ϕ	ϕ	ϕ	ϕ	ϕ	ϕ	ϕ
Feature 2	ψ_1	ψ_1	ψ_1	ψ_1	ψ_1	ψ_1	ψ_1
Feature 3	ψ_2	$\psi_{2^{2/3}}$	$\psi_{2^{1/2}}$	$\psi_{2^{2/5}}$	$\psi_{2^{1/3}}$	$\psi_{2^{2/7}}$	$\psi_{2^{1/4}}$
Feature 4	ψ_4	$\psi_{2^{4/3}}$	ψ_2	$\psi_{2^{4/5}}$	$\psi_{2^{2/3}}$	$\psi_{2^{4/7}}$	$\psi_{2^{1/2}}$
Fearure 5	ψ_8	ψ_4	$\psi_{2^{3/2}}$	$\psi_{2^{6/5}}$	ψ_2	$\psi_{2^{6/7}}$	$\psi_{2^{3/4}}$
Feature 6		$\psi_{2^{8/3}}$	ψ_4	$\psi_{2^{8/5}}$	$\psi_{2^{4/3}}$	$\psi_{2^{8/7}}$	ψ_2
Feature 7			$\psi_{2^{5/2}}$	ψ_4	$\psi_{2^{5/3}}$	$\psi_{2^{10/7}}$	$\psi_{2^{5/4}}$
Feature 8			ψ_8	$\psi_{2^{12/5}}$	ψ_4	$\psi_{2^{12/7}}$	$\psi_{2^{3/2}}$
Feature 9				$\psi_{2^{14/5}}$	$\psi_{2^{7/3}}$	ψ_4	$\psi_{2^{7/4}}$
Feature 10					$\psi_{2^{8/3}}$	$\psi_{2^{16/7}}$	ψ_4
Feature 11					ψ_8	$\psi_{2^{18/7}}$	$\psi_{2^{9/4}}$
Feature 12						$\psi_{2^{20/7}}$	$\psi_{2^{5/2}}$
Feature 13							$\psi_{2^{11/4}}$
Feature 14							ψ_8

Once visual features are clustered, annotated regions can be described by the *bag of visual words* contained in the region: these are the words corresponding to the feature vector for every pixel in the form of a histogram. Since regions are now described in terms of the histogram of visual words, maximum dimensionality is the number of visual words. The dimensionality of the feature space is reduced, as annotated regions are described by a single vector whereas they were sometimes described by thousands of features.

3   Results

In this section, the techniques described in Section 2 are applied to the data set. First, the entire database is analyzed and visual features are extracted with seven scale progression schemes, using a number of voices per octave ranging from 1 to 4 in steps of 0.5 voices, with scale values $1 \leq s \leq 8$. For each scale progression, a wavelet transform is computed on the complete image and only values for pixels contained in annotated regions of interest are included in the feature space. This produces a set of 7 feature spaces with a number of features ranging from 5 to 14 features, as shown in Table 1. Then, each of the feature spaces is clustered using k–means with 5 different numbers of clusters, varying from 5 to 25. This is done three times, using Euclidean distance, Euclidean distance in normalized feature space and Mahalanobis distance. Once the feature spaces are clustered, the images are analyzed again to assign visual words to each pixel of the annotated regions (see Figure 3) This provides a set of 105 feature sets as a result of using 7 scale progressions, 5 vocabulary sizes for each progression, and 3 distance measures for each vocabulary.

For performance evaluation, early precision (P1, precision of the first image) is calculated using a leave–one–patient–out cross–validation in order to avoid

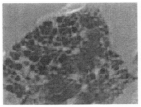

Fig. 3. Cropped example of a region described by visual words (in color)

that tissue of a patient influences classification of other tissue types of the same patient, which can create a bias. Results show that the Euclidean distance in the original feature space outperforms the other two distance measures and that 25 visual words perform better than a smaller number of visual words. Figure 4(a) shows the results for class–specific P1 for various scale progressions, which is equivalent to a kNN classification with $k = 1$. No clear trend can be extracted from the disease–wise performance, whereas in terms of geometric and arithmetic means the only trend is a slight decay with an increasing number of voices. It was initially thought that a relatively small number of words would be best for classifying the few classes. The trend for all scale progressions and distance measures shows that performance improves with a larger number of visual words. The experiment was then extended increasing the size of the visual vocabulary. The results are calculated for Euclidean distance and 1 voice scale progression, since both are simpler and perform better than other configurations. Figure 4(b) shows P1 for visual words varying from 5 to 500 for every tissue type in the database together with the geometric mean. P5 and P10, as well as precision at the number of relevant regions in the database were also calculated for the best performing configurations, showing a similar decrease in performance for all measures, without any remarkable difference (see Table 2).

Table 2. Early and late precision percentage values for 125 and 150 visual words

	P1 (%)	P5 (%)	P10 (%)	PN_r (%)
125 visual words	61.5	59.18	57.4	41.3
150 visual words	61.3	59	56.6	41.1

4   Discussion

Results in Section 3 show that the scale progression does not have a strong impact on the system performance; there is a slowly decaying trend that might be due to the fact that 1 voice (dyadic scale) already has enough information for representing the texture patterns. Therefore, a larger number of voices may add noise rather than valuable band–pass information (see Figure 4(a)).

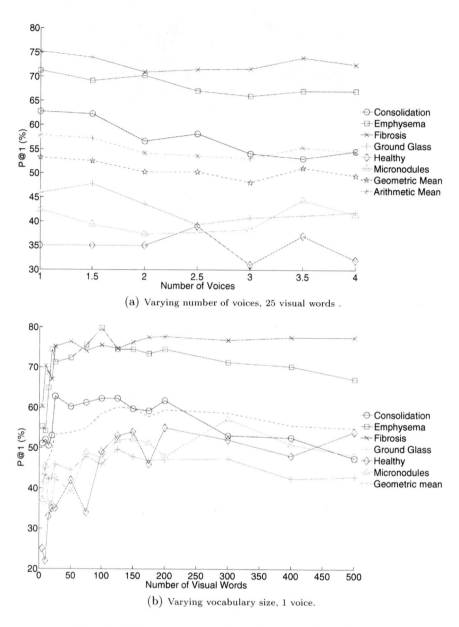

(a) Varying number of voices, 25 visual words .

(b) Varying vocabulary size, 1 voice.

Fig. 4. P1 for a varying number of voices and visual words

By extending the experiment, the optimal number of visual words is found to be between 100 and 300, where performance remains fairly stable. This can be explained by the fact that some of the 6 tissue classes may contain several visually distinct patterns that can better be described with a larger number of clusters. The slow decrease when creating even more visual words may be

Table 3. Confusion matrix of the first retrieved regions for dyadic scale progression and 125 visual words using Euclidean distance

Actual class	Predicted class					
	Consolidation	Emphysema	Fibrosis	Ground glass	Healthy	Micronodules
Consolidation	**62.2**	0	26.0	9.2	1	1.5
Emphysema	0	**74.5**	6.4	2.1	8.5	8.5
Fibrosis	5.3	1.1	**74.6**	13.3	0.5	4.8
Ground glass	4	0.7	24.4	**49.6**	0.9	20.4
Healthy	1	9	19	1	**53**	17
Micronodules	1	0.3	10.4	26.6	9.4	**51.9**

due to splitting useful clusters and thus removing important class information. Figure 4(b) shows that the increase in overall performance is mostly due to an improvement in precision for healthy tissue. This confirms the hypothesis that for an accurate description of strongly varying patterns such as healthy tissue a sufficiently high number of visual words is required.

The system presented in this work performs well in a harder scenario compared to past work using similar visual features on basically the same database [5] Whereas previous work included 99 clinical parameters in the classification, the work presented in this paper uses only visual information. The inclusion of the clinical context in retrieval can increase performance in the order of 8% [4]. Whereas leave–one–ROI–out assures that the retrieved cases do not contain the query item itself, the cross–validation scheme used for this work assures that the retrieved cases do not contain any result from the same patient. This is clearly a more realistic, yet more difficult, scenario that is unfortunately often neglected in the literature [8,12]. Particularly with a k–NN classifier, tissue samples of the same patient can often lead to seemingly better results as tissue of the same patient and with the same abnormalities is clearly much more visually similar than that of other patients.

Another common mistake is to not include healthy tissue into the retrieval database, which is the first question of a clinician in any case. Healthy tissue has a very large variety between patients, whereas pathologic tissue usually has much clearer class boundaries. In this work, healthy tissue is included as a class to be regarded, providing a more useful tool that can not only classify abnormal patterns, but detect these abnormalities from healthy parts. In Table 3 it can be seen that considering only a 2–class database (healthy and non–healthy tissue) the system would have P1 of 94.5%, with less than 3% false negatives.

5   Conclusions and Future Work

In this paper, a content–based image retrieval system using multi–scale visual words to describe lung texture associated with ILDs is proposed. The database used for evaluation is taken from clinical routine and annotated by two emergency radiologists. Results show a good performance of the system when the correct parameters are chosen for scale progression, distance metric used and particularly the number of visual words. The system is valuable for supplying clinicians with visually similar regions of interest and also visually similar cases with proven pathologies that can support diagnosis. The implemented techniques can be seen as useful in other clinical areas where texture analysis is required for diagnosis such as for analyzing liver images. The results show a limited impact of the number of bandpass wavelet filters but highlight the importance of a sufficiently high number of visual words to describe tissue types with high intra–class variations such as healthy tissue.

Further work is foreseen that may improve the system performance. Using the full 3D data for the generation of the visual words is foreseen as some tissue types can profit from available 3D information, for example to disambiguate between vessels in the lung and nodules that look very similar in 2D slices. This requires a clinical protocol with small inter slice distance, though. The current system only concentrates on visual analysis whereas the inclusion of clinical data such as age and smoking history has shown to increase classification performance in the past [4].

Acknowledgments. This work was supported by the Swiss National Science Foundation (FNS) in the MANY project (grant 205321–130046) as well as the EU 7th Framework Program in the context of the Khresmoi project (FP7–257528).

References

1. Aisen, A.M., Broderick, L.S., Winer-Muram, H., Brodley, C.E., Kak, A.C., Pavlopoulou, C., Dy, J., Shyu, C.R., Marchiori, A.: Automated storage and retrieval of thin–section CT images to assist diagnosis: System description and preliminary assessment. RAD 228(1), 265–270 (2003)
2. Aziz, Z.A., Wells, A.U., Hansell, D.M., Bain, G.A., Copley, S.J., Desai, S.R., Ellis, S.M., Gleeson, F.V., Grubnic, S., Nicholson, A.G., Padley, S.P., Pointon, K.S., Reynolds, J.H., Robertson, R.J., Rubens, M.B.: HRCT diagnosis of diffuse parenchymal lung disease: inter–observer variation. Thorax 59(6), 506–511 (2004)
3. Dance, C., Willamowski, J., Fan, L., Bray, C., Csurka, G.: Visual categorization with bags of keypoints. In: ECCV International Workshop on Statistical Learning in Computer Vision (2004)
4. Depeursinge, A., Iavindrasana, J., Cohen, G., Platon, A., Poletti, P.A., Müller, H.: Lung tissue classification in HRCT data integrating the clinical context. In: 21th IEEE Symposium on Computer-Based Medical Systems (CBMS), Jyväskylä, Finland, pp. 542–547 (June 2008)
5. Depeursinge, A., Sage, D., Hidki, A., Platon, A., Poletti, P.A., Unser, M., Müller, H.: Lung tissue classification using Wavelet frames. In: 29th Annual International Conference of the IEEE Engineering in Medicine and Biology Society, EMBS 2007, pp. 6259–6262. IEEE Computer Society, Lyon (2007)
6. Depeursinge, A., Van De Ville, D., Unser, M., Müller, H.: Lung tissue analysis using isotropic polyharmonic B–spline wavelets. In: MICCAI 2008 Workshop on Pulmonary Image Analysis, New York, USA, pp. 125–134 (September 2008)

7. Depeursinge, A., Vargas, A., Platon, A., Geissbuhler, A., Poletti, P.A., Müller, H.: Building a reference multimedia database for interstitial lung diseases. Computerized Medical Imaging and Graphics (submitted)
8. Dy, J., Brodley, C., Kak, A., Broderick, L., Aisen, A.: Unsupervised feature selection applied to content-based retrieval of lung images. IEEE Transactions on Pattern Analysis and Machine Intelligence 25(3), 373–378 (2003)
9. Fetita, C.I., Chang-Chien, K.-C., Brillet, P.-Y., Prêteux, F., Grenier, P.: Diffuse Parenchymal Lung Diseases: 3D Automated Detection in MDCT. In: Ayache, N., Ourselin, S., Maeder, A. (eds.) MICCAI 2007, Part I. LNCS, vol. 4791, pp. 825–833. Springer, Heidelberg (2007)
10. Gangeh, M.J., Sørensen, L., Shaker, S.B., Kamel, M.S., de Bruijne, M., Loog, M.: A Texton-Based Approach for the Classification of Lung Parenchyma in CT Images. In: Jiang, T., Navab, N., Pluim, J.P.W., Viergever, M.A. (eds.) MICCAI 2010. LNCS, vol. 6363, pp. 595–602. Springer, Heidelberg (2010)
11. Korfiatis, P.D., Karahaliou, A.N., Kazantzi, A.D., Kalogeropoulou, C., Costaridou, L.I.: Texture-based identification and characterization of interstitial pneumonia patterns in lung multidetector ct. IEEE Transactions on Information Technology in Biomedicine 14(3), 675–680 (2010)
12. Liu, C.T., Tai, P.L., Chen, A.Y.J., Peng, C.H., Wang, J.S.: A content–based medical teaching file assistant for CT lung image retrieval. In: Proceedings of the IEEE International Conference on Electronics, Circuits, Systems (ICECS2000), pp. 361–365. Jouneih–Kaslik, Lebanon (2000)
13. Mahalanobis, P.: On the generalised distance in statistics. Proceedings of the National Institute of Science, India 2, 49–55 (1936)
14. Müller, H., Michoux, N., Bandon, D., Geissbuhler, A.: A review of content-based image retrieval systems in medicine–clinical benefits and future directions. Internation Journal of Medical Informatics 73(1), 1–23 (2004)
15. Shyu, C.R., Brodley, C.E., Kak, A.C., Kosaka, A., Aisen, A.M., Broderick, L.S.: ASSERT: A physician–in–the–loop content–based retrieval system for HRCT image databases. Computer Vision and Image Understanding 75(1/2), 111–132 (1999)
16. Simel, D., Drummond, R.: The rational clinical examination: evidence–based clinical diagnosis. McGraw-Hill (August 2008)
17. Sivic, J., Zisserman, A.: Video google: A text retrieval approach to object matching in videos. In: IEEE International Conference on Computer Vision, vol. 2, p. 1470 (2003)
18. Uppaluri, R., Hoffman, E.A., Sonka, M., Hartley, P.G., Hunninghake, G.W., McLennan, G.: Computer recognition of regional lung disease patterns. American Journal of Respiratory and Critical Care Medicine 160(2), 648–654 (1999)
19. Yang, J., Jiang, Y.G., Hauptmann, A.G., Ngo, C.W.: Evaluating bag-of-visual-words representations in scene classification. In: Proceedings of the International Workshop on Workshop on Multimedia Information Retrieval, MIR 2007, pp. 197–206. ACM, New York (2007)
20. Zheng, Y., Greenleaf, J., Gisvold, J.: Reduction of breast biopsies with a modified self-organizing map. IEEE Transactions on Neural Networks 8(6), 1386–1396 (1997)

Histology Image Indexing Using a Non-negative Semantic Embedding

Jorge A. Vanegas, Juan C. Caicedo, Fabio A. González, and Eduardo Romero

Bioingenium Research Group
National University of Colombia
{javanegasr,jccaicedoru,fagonzalezo,edromero}@unal.edu.co

Abstract. Large on-line collections of biomedical images are becoming more common and may be a potential source of knowledge. An important unsolved issue that is actively investigated is the efficient and effective access to these repositories. A good access strategy demands an appropriate indexing of the collection. This paper presents a new method for indexing histology images using multimodal information, taking advantage of two kinds of data: visual data extracted directly from images and available text data from annotations performed by experts. The new strategy called Non-negative Semantic Embedding defines a mapping between visual an text data assuming that the latent space spanned by text annotations is good enough representation of the images semantic. Evaluation of the proposed method is carried out by comparing it with other strategies, showing a remarkable image search improvement since the proposed approach effectively exploits the image semantic relationships.

1   Introduction

Digital microscopy is currently an important tool to support the decision making process in clinical and research environments. Instead of moving the glass around, specialists can share a digitized sample, allowing other physicians to navigate the slide by the use of a virtual microscope [1]. The digitization process results in very large files known as virtual slides, or alternatively, many different microphotographs taken from the same slide. A large database of these micropictures can support training, educational and research activities. Likewise, the great amount of information contained in these image collections and their accompanying meta-data is a potential resource to support the specialist's decision-making processes. However, accessing and retrieving images from a large collection is a high time consuming task. In this work, we consider the problem of retrieving histological images using as query an example image. Under this setup, the system relies mainly on processing the visual contents to find relevant images. Yet, matching visual similarities does not necessarily lead to meaningful results, a problem known as the semantic gap [2].

To overcome this problem, semantic indexing has been proposed by exploiting additional information resources, such as image meta-data and accompanying text. The main problem is that collecting that data is an arduous process which

H. Müller et al. (Eds.): MCBR-CDS 2011, LNCS 7075, pp. 80–91, 2012.

makes it very difficult to ensure an adequate annotation for each image in a very extensive collection. Therefore, we are in a situation in which we have lots of images available and only a small portion of them is actually annotated. So, the challenge is to design a method that takes advantage of the semantic information extracted from annotated images to improve the search process in the entire collection.

In this paper we address the problem of indexing histological images, using the multimodal information drawn from two kinds of data: images and the available text from annotations. We introduce a new strategy to find the relationships between these two data modalities, the Non-negative Semantic Embedding, which defines a mapping between visual and text data. Using this approach, the system is able to project new images to the space defined by the semantic annotations. This work presents two main contributions: first, a new method to index images using multimodal information, and second, an experimental evaluation of histological images which defines the role of semantic data in this particular field. The rest of this paper is organized as follows: Section 2 discusses the related work; Section 3 presents the structure of the histology image collection used in this work; Section 4 introduces the proposed method called Non-negative Semantic Embedding; Section 5 presents the experimental evaluation; and, finally, Section 6 presents some concluding remarks.

2   Related Work

Access to histological image repositories using a query-by-example (QBE) approach based on low-level visual features has been investigated in different works [3,4]. Subsequent works [5,6] showed that the semantic analysis and automatic classification of histology images can be used to improve the response of a image retrieval system.

The ImageCLEFmed campaign organizes an experimental evaluation of medical image retrieval, providing access to a large dataset with multimodal data (images and text) and with an important emphasis on radiological images. The 2010 version of this event showed that the combination of the visual and text modalities has important benefits in retrieval performance [7]. Similarly, our approach intends to exploit multimodal data in histology image collections for building a semantic index that supports QBE.

Recently, a strategy for multimodal indexing of images was proposed to improve the performance of a retrieval system that works under the QBE paradigm [8]. This strategy models latent factors using non-negative matrix factorization to find meaningful relationships between visual and text data. We build on top of these ideas to propose a novel algorithm, non-negative semantic embedding, which is used to build a semantic index of histology images.

3   Histology Images

Images in this work have been collected to build an atlas of histology for the study of the four fundamental tissues. The collection comprises 20,000 images of these tissues, from different biological systems and organs, which were stained using hematoxiline/eosine and Immunohistochemical techniques. The collection includes photographs of histology slides acquired with a digital camera coupled to a microscope, using different magnification factors to focus important biological structures. The main use of this collection is for educational and research purposes, and is accessible via Internet at www.informed.unal.edu.co.

From this large collection, a subset of 2,641 images was selected for this work. Each of these images was annotated by an expert, indicating the biological system and organs that can be observed. The total number of different categories is 46, after a standardization of the vocabulary used to describe the semantic contents. The list of terms includes circulatory system, hearth, lymphatic system and thymus, among others. Usually, images have just one category attached to it, but in several cases images can have more than one category . Figure 1 shows example images with the corresponding associated terms.

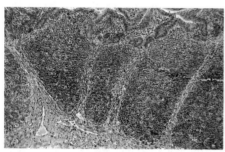 

Category: Lymphatic system. Lymphatic structure of the digestive tract.
Category: Digestive system. Appendix.

Category: Female reproductive system. Ovarian.

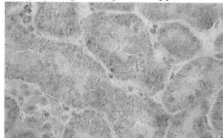 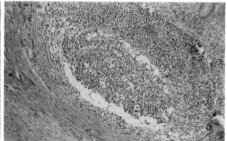

Category: Urinary system. Kidney.

Category: Lymphatic system. Lymphatic structure of the digestive tract.
Category: Digestive system. Ileus.

Fig. 1. Example images with associated terms

4   Matrix Factorization for Multimodal Indexing

Assume a discrete image representation for the visual contents using, for instance, a bag of features, i.e., a random collection of parts of the image contents whose distribution probability is approached by a simple frequentist approach, that is to say, the frequency of these patches is determined within the whole image collection. Similarly, assume a text representation for annotations, also approached by term frequencies in the attached annotations and also using a bag of words. Then, we can build two matrices to describe the occurrence of visual and textual features in the image collection. Let $X_v \in \mathbb{R}^{n \times l}$ be the matrix of visual features, where n is the number of visual features and l is the total number of images in the collection. Let $X_t \in \mathbb{R}^{m \times l}$ be the matrix of text term frequencies, where m is the number of terms or keywords.

Our goal is to uncover the underlying structure of these matrices to build an effective index for image search. The proposed strategy is based on the Non-negative Matrix Factorization (NMF) algorithm, to find a linear representation of the data, which must be non negative [9].

4.1   Non-negative Matrix Factorization

The main purpose of NMF is to find an approximation of the matrix X, in terms of two smaller matrices as follows:

$$X = WH . \tag{1}$$

where $X \in \mathbb{R}^{p \times l}$, $W \in \mathbb{R}^{p \times r}$, $H \in \mathbb{R}^{r \times l}$, p is the total number of available features, l is the number of images in the collection, and r is the rank of the decomposition. The matrix W is known as the basis matrix and H is known as the encoding matrix. This factorization is found by solving the associated optimization problem that corresponds to minimize the squared error of the decomposition or the Kullback Leibler divergence between the original matrix and the reconstructed one.

We use the divergence criterion in this work, following the Lee and Seung's approach [10] to find the decomposition with two multiplicative updating rules for W and H, respectively:

Updating rule for W

$$W_{ia} \leftarrow W_{ia} \frac{\sum_\mu H_{a\mu} X_{i\mu}/(WH)_{i\mu}}{\sum H_{av}} . \tag{2}$$

Updating rule for H

$$H_{a\mu} \leftarrow H_{a\mu} \frac{\sum_i W_{ia} X_{i\mu}/(WH)_{i\mu}}{\sum_k W_{ka}} . \tag{3}$$

4.2   Multimodal Indexing via NMF

González et al. [8] proposed two ways of using NMF for multimodal indexing of images. The first strategy, the NMF-mixed, builds a multimodal matrix $X = \left[\alpha X_v^T \ (1 - \alpha) X_t^T \right]^T$ with $\alpha \in [0, 1]$, that contains the visual and text data. Matrices are mixed after normalization of each vector to have L2-norm $= 1$. This matrix is then decomposed using NMF to model a set of latent factors (columns of the matrix W) that have features of both modalities.

The second strategy, the NMF-asymmetric, follows a two step process, starting with the decomposition of the text matrix. After that, the obtained encoding for images (columns of the matrix H) is fixed to find a second matrix of latent factors for visual contents. The goal is to find a semantic image representation based on the relationships between text terms only, and then find a function to project visual features to the same latent space. The two steps of the NMF-asymmetric strategy are as follows:

$$X_t = W_t H \ . \tag{4}$$

$$X_v = W_v H \ . \tag{5}$$

where W_t is the matrix of latent factors for text, W_v is the matrix of latent factors for visual features and H is the common representation for both modalities. The matrix W_v is obtained by running the multiplicative update for W only (Equation 2) since the matrix H is fixed.

4.3   Non-negative Semantic Embedding

The methods described in the previous Subsection are oriented to model latent factors for multimodal data, that is, to find the hidden structure of the collection, which is assumed to be common between the two data modalities. We propose a simplified strategy that extends the NMF-asymmetric algorithm to a setting in which the semantic encoding is already known.

We assume that the space spanned by text terms is a good enough representation of image semantics, and we use it to index and represent all images in the collection. Then, we want to find a way to embed visual features in this semantic space, to index images with or without annotations. We formulate this problem as finding a linear transformation of the visual data imposing a non negativity constraint on the solution, as follows:

$$X_v = W X_t; W \geq 0 \ . \tag{6}$$

where $W \in \mathbb{R}^{n \times m}$, n the number of visual features and m the number of text terms. The non-negativity constraint in this case enforces an additive reconstruction of visual features, since vectors in the matrix W can be thought of as parts of images that are combined according to the presence of associated labels. Notice that the encoding matrix is the matrix of text annotations, and the vectors in W can also be directly interpreted as the visual features related to each text term.

Finding the matrix W when X_v and X_t are known is a convex problem under the divergence or euclidean criteria in the related minimization problem. However, we approximate the solution to this problem using the NMF updating rule for the matrix W. Instead of requiring a global optimum for this problem, which might result in overfitting to the training data, we accept a good approximate solution obtained from the Lee and Seung's approach. The updating rule usually converges to a local minimum, but it may result in a better generalization with some robustness to intrinsic noise in the training data.

We call this approach the Non-negative Semantic Embedding (NSE) taking into account that the semantic space is known in the problem and the resulting solution to embed image features on it is non-negative. It also differs from the Gonzalez et al. [8] approach in the sense that no latent factors are herein modeled, but instead, a semantic space is assumed from the given text terms space.

4.4   Image Indexing and Search

The indexing methods described above require a training phase to learn a mapping of images to the semantic or latent space. The result of that training phase is a matrix W, which contains the basis of the indexing space and serves as linear transformation to project new data. So, when new images are obtained without text annotations, either to be included in the collection or as queries, we can find the semantic representation for this image using visual features only.

Let y be the visual features of a new image, unseen during training. To embed this image in the semantic space, the following equation needs to be solved:

$$y = Wh \ . \tag{7}$$

where W is the basis of the semantic space and h is the semantic representation of the new image. The image y will be embedded into the semantic space by finding the vector $h \geq 0$ that satisfies the equation. This is done by using the multiplicative updating rule for h while keeping the matrix W fixed.

This strategy applies for both, the NMF-based algorithms using latent factors and the NSE. So, now that we can represent images in the collection and query images in same space, the problem of finding relevant results reduces to the problem of matching images with similar representations. To do so, we employ the dot product as similarity measure, which gives a notion of the extent to which two images share similar components in the latent or semantic space. Finally, results are ranked in decreasing order of similarity and delivered to the user.

5   Experiments and Results

5.1   Experimental Setup

We conducted retrieval experiments under the query-by-example paradigm to evaluate the proposed methods. A set of 100 images were randomly selected as queries from the database of 2,641 images used in this study. The remaining 2,541 images were used as the target collection to find relevant images.

We performed automatic experiments by sending a query to the system and evaluating the relevance of the results. A ranked image in the results list is considered relevant if it shares at least one keyword with the query. For this experiment, the evaluation was done using traditional measures of Image Retrieval, including Mean Average Precision (MAP) and the Recall-Precision plots.

Image Features. We build a bag-of-features representation for the set of histological images, as it has been found to be an effective representation for microscopy image analysis [11,12]. We start by extracting patches of 8x8 pixels from a set of training images with an overlap of 4 pixels along the x and y axes. The DCT (Discrete Cosine Transform) transform is applied in each of the 3 RGB channels to extract the largest 21 coefficients and their associated functions. A k-means clustering is applied to build a dictionary of 500 visual terms. This bag-of-features configuration have shown good results with similar types of histology images [11,12].

Once the vocabulary has been built, every image in the collection goes through the patch extraction process. Each patch from an image is linked to one visual term in the dictionary using a nearest neighbor criterion. Finally, the histogram of frequencies is constructed for each image. We experimentally found that 500 visual terms was enough to achieve a good performance, and that larger dictionaries do not provide significant improvements, but just more computational load.

Text Annotations. In this data set of histology images, text annotations are clean and clearly defined terms from a technical vocabulary. Since the annotation process followed a systematic revision, there is no need to build a vector space model or to account for term frequencies. We build semantic vectors following a boolean approach, assigning 1 to the terms attached to an image and 0 otherwise. This leads to 46-dimensional binary vectors, which serve to build the text matrix.

5.2   Experiments

Visual Search. As a first experiment we retrieved histology images using only visual information as a baseline to assess improvements of other methods. The visual descriptors are based on the bag-of-features strategy, so images are represented by histograms of the occurrence of visual patterns in a dictionary. Direct visual matching is done by calculating the level of similarity between images using the histogram intersection similarity measure [13], as follows:

$$K_{HI}(x, y) = \sum_{i=1}^{n} min\{x_i, y_i\} . \tag{8}$$

where x and y are images and x_i, y_i are the i-th occurrence of the visual feature in these images, respectively. Using direct matching only, the system achieves a performance up to 0.210 in terms of MAP. An additional experiment using visual features was conducted to determine if latent factors learned only from visual data can help to improve over this baseline.

We applied an NMF decomposition on the visual matrix X_v, using different numbers of latent factors. In the best performing case, this strategy reaches a value of 0.172 of MAP. This suggest that the dimensionality reduction made by NMF leads to a loss of discriminative power of visual descriptors instead of helping to identify semantic patterns in the collection.

Multimodal Search. Multimodal search aims to introduce text information during the search process, even though in our case queries are expressed using no keywords but example images. To this end, we employ three different algorithms based on NMF: NMF-mixed, NMF-asymmetric and NSE, and compare their performance against visual search. For the first algorithm (NMF-mixed) the construction of a multimodal matrix was done by setting $\alpha=0.5$ to give the same importance to visual and text data.

For the NMF-mixed and NMF-asymmetric algorithms we performed several experiments using different sizes of the latent semantic space, to experimentally determine an appropriate number of latent factors. In contrast, the NSE algorithm does not need to set this parameter. Figure 2 shows the result of exploring the number of latent factors for these algorithms, including reference lines for the visual search and NSE. It can be seen that the response of NSE outperforms by a large margin the response of all other strategies. In addition, the Figure shows that besides NSE, only the NMF-asymmetric strategy is able to improve over the visual baseline.

The loss in performance of using NMF-mixed and NMF-visual can be explained by the difficulty of these strategies to find meaningful latent factors from the given input data. As was mentioned before, the NMF-visual fails to find semantic patterns in the collection leading to a decreasing of the discriminatory power of the full visual representation. The NMF-mixed shows a better achievement in this setting, which improves over NMF-visual due to the presence of text terms in the multimodal matrix. Still, the semantic patterns are not correctly modeled by multimodal factors because they have to deal with the reconstruction of visual features as well.

The two strategies that improve over the direct visual matching are NSE and NMF-asymmetric, which concentrate on exploiting text information as the semantic reference data. The NMF-asymmetric, in particular, decomposes the text matrix in an attempt to find meaningful relationships between text terms to build semantic latent factors. In our experiments, we decompose the matrix of 46 terms in 10, 20, 30 and 40 latent factors, from which the second choice presented the best performance. However, the improvement of NMF-asymmetric is modest with respect to NSE, indicating that latent factors are not a fundamental modelling aspect for this dataset.

Another consideration of the modest improvement of NMF-asymmetric may be found in the two steps algorithm. When the first decomposition is done, an approximation error is generated since perfect reconstruction is not required. Then, the second decomposition builds on top of it to construct latent factors for visual features, introducing its own approximation error as well. The NSE algorithm simplifies the approach by learning a unique matrix that correlates visual and text data directly.

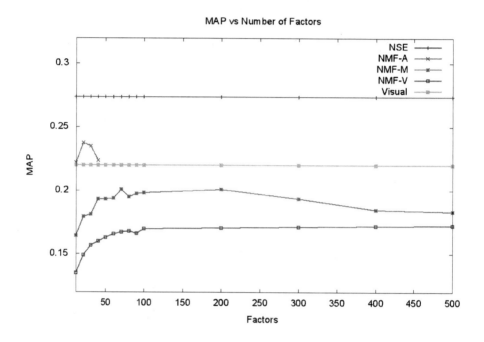

Fig. 2. Latent factors vs. MAP

Another way to measure the performance of the evaluated algorithms is using the Recall vs. Precision graph. For this evaluation, we selected the number of latent factors that provided the best performance in the previous evaluation. We plot the interpolated Recall-Precision graph, in which one can observe the differences in precision along the retrieval process, i.e., while all relevant images are being retrieved to the user. Figure 3 shows the result of this evaluation, and reveals that the direct visual matching presents very good results in early precision, but also falls very fast as long as more relevant images are required.

The second best performance in the first positions of the results page is given by the NMF-visual approach, according to the Recall-Precision graph. What it actually means, is that nearest neighbors under a visual similarity measure are very likely to be relevant in a histology image collection. This contrast with all other semantic indexing approaches (NMF-M, NMF-A and NSE) which are modeling structural patterns in the whole collection, rather than exploiting the local similarities of the dataset. Nevertheless, the performance of NSE and

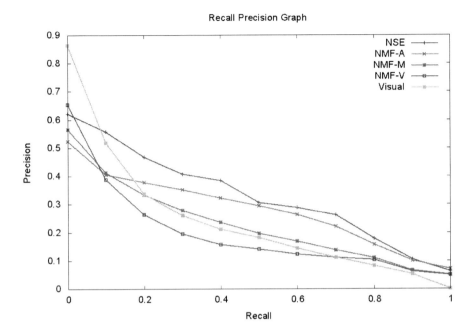

Fig. 3. Recall vs. Precision Graph

NMF-asymmetric showed a more consistent and sustained higher precision as long as the user explores more images in the results page.

The good performance obtained in early position with the direct visual matching method are due to the fact that in this collection several histology images come from sections that belong to the same histological plate or represent the same region with different zoom levels. Therefore, in this case, visual relationships directly denote semantic relations. But this behavior only occurs in the first retrieved images. In the Recall-Precision graph shows that values of recall greater than 10% (that in this collection represents an average of 20 images) the accuracy of the visual methods falls dramatically, while the NSE shows a more stable behavior.

Table 1 summarizes the findings of our experimental results, presenting the number of required latent factors, MAP measures, relative improvement in terms

Table 1. Performance measures for all evaluated strategies

Method	Latent Factors	MAP	Improvement	P@10
Visual Matching	N/A	0.210	N/A	0.650
NMF-Visual	300	0.172	-18.8%	0.088
NMF-Mixed	70	0.201	-4.3%	0.252
NMF-Asymmetric	20	0.235	+11.9%	0.208
NSE	N/A	0.273	+30.0%	0.265

of MAP w.r.t. the visual baseline and the precision at the first 10 results (P@10). The Table shows that NMF-asymmetric and NSE provide a significant improvement in terms of MAP compared with only visual retrieval. It also confirms the dominant position of visual search in terms of early precision.

6 Conclusions

We presented a method for image indexing that combines visual and text information using a variation of non-negative matrix factorization. Text annotations provide a semantic representation space where the visual content of images is embedded using NMF.

The experimental evaluation demonstrates an increase in retrieval performance. The effectiveness of this method may be explained by the fact that it efficiently exploits the semantic information contained in text annotations. But we must take into account that this methodology is mainly applicable in cases where we have a controlled annotation process.

An important characteristic of the proposed method is that it naturally deals with different types of data: non-annotated images, images with multiple annotations, text queries, etc. This is accomplished by mapping both images and queries (textual and visual) to a common semantic representation space. Another important advantage is that the method could be efficiently implemented by integrating the last developments on on-line matrix factorization (such as the method proposed by Mairal et al. [14]) which can deal with large amounts of data.

References

1. Roa-Peña, L., Gómez, F., Romero, E.: An experimental study of pathologist's navigation patterns in virtual microscopy. Diagnostic Pathology 5(1), 71 (2010)
2. Smeulders, A.W.M., Worring, M., Santini, S., Gupta, A., Jain, R.: Content-based image retrieval at the end of the early years. IEEE Transactions on Pattern Analysis and Machine Intelligence 22(12), 1349–1380 (2000)
3. Zheng, L., Wetzel, A.W., Gilbertson, J., Becich, M.J.: Design and analysis of a content-based pathology image retrieval system. IEEE Transactions on Information Technology in Biomedicine 7(4), 249–255 (2003)
4. Caicedo, J.C., González, F.A., Triana, E., Romero, E.: Design of a Medical Image Database with Content-Based Retrieval Capabilities. In: Mery, D., Rueda, L. (eds.) PSIVT 2007. LNCS, vol. 4872, pp. 919–931. Springer, Heidelberg (2007)
5. Yu, F., Ip, H.: Semantic content analysis and annotation of histological images. Computers in Biology and Medicine 38(6), 635–649 (2008)
6. Caicedo, J.C., Cruz, A., Gonzalez, F.A.: Histopathology Image Classification using Bag of Features and Kernel Functions. In: Combi, C., Shahar, Y., Abu-Hanna, A. (eds.) AIME 2009. LNCS, vol. 5651, pp. 126–135. Springer, Heidelberg (2009)
7. Müller, H., Clough, P., Deselaers, T., Caputo, B.: ImageCLEF: Experimental Evaluation in Visual Information Retrieval. Springer, Heidelberg (2010)

8. González, F.A., Caicedo, J.C., Nasraoui, O., Ben-Abdallah, J.: NMF-based multimodal image indexing for querying by visual example. In: ACM International Conference On Image And Video Retrieval, pp. 366–373. ACM Press, New York (2010)
9. Liu, W., Zheng, N., Lu, X.: Non-negative matrix factorization for visual coding. In: 2003 IEEE International Conference on Acoustics Speech and Signal Processing 2003 Proceedings, vol. 3, pp. III–293–6. IEEE (2003)
10. Lee, D.D., Seung, H.S.: New Algorithms for Non-Negative Matrix Factorization in Applications to Blind Source Separation. In: 2006 IEEE International Conference on Acoustics Speed and Signal Processing Proceedings, vol. 13(1), pp. V–621–V–624 (2001)
11. Díaz, G., Romero, E.: Histopathological Image Classification using Stain Component Features on a pLSA Model. In: Bloch, I., Cesar Jr., R.M. (eds.) CIARP 2010. LNCS, vol. 6419, pp. 55–62. Springer, Heidelberg (2010)
12. Cruz-Roa, A., Caicedo, J.C., González, F.A.: Visual Pattern Analysis in Histopathology Images Using Bag of Features. In: Bayro-Corrochano, E., Eklundh, J.-O. (eds.) CIARP 2009. LNCS, vol. 5856, pp. 521–528. Springer, Heidelberg (2009)
13. Swain, M.J., Ballard, D.H.: Color indexing. International Journal of Computer Vision 7, 11–32 (1991)
14. Mairal, J., Bach, F., Ponce, J., Sapiro, G.: Online learning for matrix factorization and sparse coding. J. Mach. Learn. Res. 11, 19–60 (2010)

A Discriminative Distance Learning–Based CBIR Framework for Characterization of Indeterminate Liver Lesions

María Jimena Costa[1], Alexey Tsymbal[1],
Matthias Hammon[3], Alexander Cavallaro[3],
Michael Sühling[1], Sascha Seifert[1], and Dorin Comaniciu[2]

[1] Siemens Corporate Technology, Erlangen, Germany
[2] Siemens Corporate Research, Princeton, USA
[3] University Hospital Erlangen

Abstract. In this paper we propose a novel learning–based CBIR method for fast content–based retrieval of similar 3D images based on the intrinsic Random Forest (RF) similarity. Furthermore, we allow the combination of flexible user–defined semantics (in the form of retrieval contexts and high–level concepts) and appearance–based (low–level) features in order to yield search results that are both meaningful to the user and relevant in the given clinical case. Due to the complexity and clinical relevance of the domain, we have chosen to apply the framework to the retrieval of similar 3D CT hepatic pathologies, where search results based solely on similarity of low–level features would be rarely clinically meaningful. The impact of high–level concepts on the quality and relevance of the retrieval results has been measured and is discussed for three different set–ups. A comparison study with the commonly used canonical Euclidean distance is presented and discussed as well.

Keywords: CBIR, distance learning, liver lesion, random forest, context–specific retrieval.

1   Introduction

Possible causes for liver lesions are varied; they may originate from both malignant (e.g. metastases) or benign (e.g. cysts or hemangiomas) pathologies [1]. The characterization of these abnormal masses constitutes an essential task on which both diagnosis and the eventual treatment of the patient are based. Factors such as size, number, shape, margin definition or enhancement pattern in different contrast phases have a strong impact on the subsequent decisions.

A contrast agent is often administered to the patient, and several consecutive CT scans are then acquired within some minutes. The degree of contrast agent present in the liver at each acquisition's time (i.e., the contrast agent phase) can have a severe impact on the appearance of abnormal masses. While the same lesion may look radically different in each of these consecutive CT images, two lesions originated by very different pathologies that pose different risks for the patient may have almost identical appearance in some of them (see Figure 1).

H. Müller et al. (Eds.): MCBR-CDS 2011, LNCS 7075, pp. 92–104, 2012.
© Springer-Verlag Berlin Heidelberg 2012

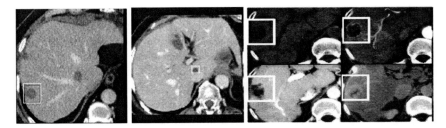

Fig. 1. While the two leftmost lesions have similar appearance, the leftmost is malignant (metastasis) while the one in the center is benign (cyst). The rightmost image shows the varying appearance of the same lesion in 4 different contrast agent phases. In such scenarios, a search relying solely on image–based similarity criteria (i.e. appearance) would probably fail to yield meaningful results.

While similar lesion retrieval can be a powerful decision support tool, searches based solely on appearance criteria would seriously limit the clinical meaningfulness of the retrieved results. Much more relevant results can be produced if the retrieval system is able to respond to user requests such as "Find lesions with comparable benignancy, given that the input lesion is in Portal Venous phase" or "Find lesions that are likely to be of the same type given that the input lesion is located in liver segment 4a and that it has a defined margin". The possibility to specify a desired *context* (i.e. similarity criterion, e.g. "comparable benignancy" or "similar margin definition") as well as additional *semantic* (i.e. *high–level*) features or concepts (e.g. "given that the input lesion is in Portal Venous phase" or "given that the input lesion is located in liver segment 4a") play an important role when tailoring the results to the user expectations.

Towards this goal, we introduce a fast, flexible, learning–based semantic CBIR framework that can accommodate combinations of appearance–based and user–defined semantic similarity criteria in order to retrieve the lesions most relevant in the given clinical context. Furthermore, once the learning phase has been completed, the method does not rely on previously annotated volumes, which allows for a fast and simple addition of new volumes to the search database if needed.

The outline of our method (the application phase) is as follows:

1. **ROI selection**
   The user selects a region of interest within an image, containing a structure (e.g. liver lesion) that he wishes to use as input for the search.
2. **Specification of semantic context(s) and additional features**
   The desired relevance/similarity context(s) that are to be used for the search are specified (e.g. benignancy criterion, lesion type criterion, etc.). The user may also choose to add semantic information as input to further characterize the structure of interest (e.g. margin enhancement or location within the liver).
3. **Image–based feature computation**
   A set of appearance–based features is computed on the selected region.

4. **Distance/Similarity computation**
   A set of Random Forest classifiers associated with the chosen similarity context(s) computes distances (i.e. similarity measures) between the input lesion (characterized by the given high–level concepts and low–level features computed) and the samples contained in an image database.
5. **Ranking and presentation of results**
   Results are grouped and ordered based on the desired combination of criteria. A tailored, ranked list of retrieval results is finally produced in the form of an *html* report.

The complete workflow has been tested in three different context scenarios: lesion density, lesion benignancy, and lesion type, and the impact of additional input semantic concepts has been measured, with promising results.

2   Related Work

According to Akgül et al. [2], radiology images pose specific challenges compared with images in the consumer domain; they contain varied, rich, and often subtle features that need to be recognized in assessing image similarity. Moreover, radiology images also provide rich opportunities for CBIR: rich metadata about image semantics are often provided by radiologists, and this information is not yet being used to its fullest advantage in CBIR systems.

In a closely related work, Napel et al. [3] study how low–level computer–generated features can be combined with semantic annotations in order to improve automatic retrieval of CT images of liver lesions. Combining the features has been shown to improve the overall performance in terms of discriminating lesion types. The study has been carried out on 10 cystic and 13 metastatic volumes to evaluate one context (lesion type); 95% mean retrieval accuracy is reported. We, on the other hand, use 1103 annotated metastases and 98 cysts in our study, and test 2 additional contexts, obtaining comparable accuracy results.

Automated decision support and lesion characterization are important in particular for small indeterminate lesions where uncertainty in characterization and treatment planning may remain even after collective discussion in a multidisciplinary setting [4]. In comparison to [3], we also consider the performance with small sub–centimeter lesions separately.

A measure of image similarity can be improved through the use of classification models that categorize new instances using a training set of instances with high–level semantic annotations, which constitutes a promising attempt to close the so–called *semantic gap* between the content–based description of an image and its meaning [5]. Segal et al [6] demonstrate that high–level information extracted from imaging for liver lesions may include valuable information, and in particular may reconstruct as much as 78% of the global gene expression profiles, revealing cell proliferation, liver synthetic function, and patient prognosis.

Seifert et al. [7] propose a learning–based "search by scribble" system and apply it to the retrieval of similar liver lesions in 3D CT scans. In this case, the search context is fixed and based on pairwise similarity scores annotated for 160 hepatic lesions, with the similarity criterion left to the discretion of the annotator. Additional information about the lesion's anatomical location (i.e. the liver) is used to eliminate retrieval results that are located in other structures (e.g. in the spleen). The reported mean average precision is of 88%.

Although most statistical pattern recognition techniques are based on the Euclidean distance, one often requires non-Euclidean and non-metric similarity especially when contours, spectra or shapes are compared, for robust object matching [8]. Two non-metric distance measures based on *discriminative distance learning* has been introduced in [9]; the intrinsic Random Forest similarity and learning from equivalence constraints, with an application to anatomy surface mesh retrieval for decision support, among other applications including retrieval of clinical records, microarray gene expression and mass spectroscopy data.

The *intrinsic RF similarity* can be used for different tasks related to the classification problem. Thus, Shi et al. [10, 11] successfully use it for hierarchical clustering of tissue microarray data. First, the unlabeled original data are expanded with a synthetic class of evenly distributed instances, then an RF is trained to discriminate the original instances and the synthetic class, so that the intrinsic RF similarities can be determined and instances clustered. The resulting clusters are shown to be clinically more meaningful than the Euclidean distance based clustering with regard to post-operative patient survival. Hudak et al. [12] use the RF similarity for nearest neighbour imputation on forestry sensor data. They conclude that the RF distance based imputation is the most robust and flexible among the imputation techniques tested. It is interesting that using this similarity for the most immediate task, case retrieval and nearest neighbour classification or regression, is rather uncommon compared with its use for clustering. In one of related works, Qi et al. [13] use it for protein–protein interaction prediction, and the results compare favourably with all previously suggested methods for this task. In one imaging application, Vitanovski et al. [14] study the use of RF similarity–based regression for similar pulmonary trunk model retrieval in order to reconstruct the full surface of pulmonary trunk from incomplete sparse MRI data.

RF has been successfully applied to a number of different imaging tasks over the last several years. However, to the best of our knowledge, we are the first to consider the intrinsic RF similarity in the context of CBIR in this paper.

3 Method

3.1 Intrinsic RF Similarity

For a Random Forest trained for a certain classification problem, the proportion of the trees where two instances appear together in the same leaves can be used as a measure of similarity between them [15]. For a given forest f the similarity between two instances x_i and x_j is calculated as follows. The instances are propagated down all K trees within f and their terminal positions z in each of the trees ($z_i = (z_{i1}, , z_{iK})$ for x_i, similarly z_j for x_j) are recorded. The similarity between the two instances then equals to:

$$S(x_i, x_j) = \frac{1}{K} \sum_{k=1}^{K} I(z_{ik} = z_{jk}) \tag{1}$$

where I is the indicator function. When dissimilarity or distance is needed and not a similarity (e.g., for clustering or multi-dimensional scaling) it is normally calculated as suggested by Breiman [15]:

$$D(x_i, x_j) = \sqrt{1 - S(x_i, x_j)} \tag{2}$$

The intrinsic RF dissimilarity (equation 2) is known not to be metric [9], as the triangular inequality (equation 3) is often violated:

$$\forall x_i, x_j, x_k : D(x_i, x_j) \leq D(x_i, x_k) + D(x_k, x_j) \tag{3}$$

Several reasons motivate the choice of learning algorithm in our framework. First, RF was demonstrated to work well with and be robust to high-dimensional data with many weakly relevant, redundant and noisy features, without the need for additional data pre-processing and feature selection. Next, RF-based models are relatively fast to train and to apply comparing for example with Support Vector Machines. Then, RFs can be trained both for classification and regression problems, support supervised learning from multiple categories, and can easily handle missing values. Last but not least, they are able to provide the intrinsic RF similarity, which helps to combine the discriminative power and robustness of RFs with the transparency of case retrieval and nearest neighbour classification or regression.

Thanks to the appealing properties of RF, the RF similarity (1) can be easily calculated for different tasks, including classification and regression problems, tasks with heterogeneous feature vectors with possible missing values, and multiclass tasks.

3.2   Online RF

In order to speed up our extensive experiments, make it possible to validate more trends and make the models adaptive to learn from new samples we implement an incrementalization of RF similar to Saffari [16]. While successful lossless incrementalizations exist for many learning algorithms, most strong techniques applied in real applications are still difficult to incrementalize, and among them perhaps the most prominent example that received considerable attention lately is the ensemble of randomized decision trees, the most famous representative of which is the Random Forest algorithm [15]. A few algorithms for online ensembles of randomized decision trees have been recently proposed, and their application to vision tasks (in particular tracking) have been considered, despite the fact that they are not lossless and often require considerably more training cases than the corresponding batch technique in order to converge.

In particular, the online RF of Saffari et al. [16] trains decision trees of fixed depth and has a fixed structure which does not change with the observation of new cases once the tree depth limit is reached. We address this issue in our online RF algorithm with

- The use of primed off-line learning to speed up convergence to a reasonable accuracy.
- Different sources of randomness (including bagging modeled with Poisson distribution of instance weights, and a random sample of observed features at each node).
- Memory management to avoid exceeding a specified memory limit for the model.
- Restructuring of the trees according to observed changes in the data distribution.

The training process starts with constructing a canonical RF [15] with 100 trees. Each decision tree is subsequently refined using online training. Online learning allows the framework to be scalable to the number of training instances. For each tree, a random sample of 200 lesions with even class distribution is used for the primed training. Primed off-line training is a simple but effective technique to improve the predictive performance of the final model (see [17] for an example).

We exploit the memory management scheme proposed before for online Hoeffding trees that dynamically activates most promising nodes, for tracking feature distributions and a split attempt, and deactivates and removes the less promising ones [18]. Similar to Saffari et al. [16] and different to the Hoeffding tree, a split is simply generated after observing a certain specified number of instances (40 is the default value normally leading to best performance). For each feature, Gaussian distribution is assumed and is tracked online, and a split threshold value which maximizes the *Gini Index* value is selected.

3.3   Low–Level Liver Lesion Descriptors

Each liver lesion in our experimental setting is described by a set of low–level computer–generated imaging features as follows:

- Relative frequency histogram and four first central normalized moments on it for the Hounsfield Units (HU) distribution in the bounding box for the lesion;
- The set of eight invariant Hu moments of order up to 3 [19];
- Six invariant Zernike moments [20];
- HU histogram and the four first central moments for the whole liver;

The first three feature types describe the lesion itself, while the last one describes the image of the whole liver. Inclusion of whole liver features has been shown to always lead to an improved discrimination performance in our set of experiments.

We use 2D Hu and Zernike invariant moments [19, 20]. In order to adapt them to the 3D liver lesion, for each lesion we generate 3 orthogonal 2D cuts intersecting at the center of the lesion ROI. The invariant moments are then calculated for each cut, and the feature vector includes both the moment for each separate cut and the averaged moments.

3.4   High–Level Concepts

Each liver lesion in a CT scan has been also annotated by 2 clinical experts with a set of 20 semantic labels. Some of these descriptors correspond to or are similar to features identified in Segal et al. [6] and Napel et al. [3]. Semantic features identified in these two studies were used as inspiration during our meetings with the clinical partners.

While some features are relatively inter–observer stable (lesion margin, rim continuity), some discrepancies arise concerning mostly lesion type (i.e. diagnosis). In such cases, information from additional studies (e.g. MRI reports or biopsies) has been used to resolve differences.

The semantic features used in our studies are as follows:

- Contrast agent phase (Native, Arterial, Portal Venous, Late, N/A);
- Lesion focality (Single, Multiple, N/A);
- Lesion surrounding (Complete, Incomplete, Absent, N/A);
- Rim continuity (Continuous Bright Rim, Discontinuous Bright Rim, Continuous Dark Rim, Discontinuous Dark Rim, N/A);
- Margin (Regular, Irregular, N/A);
- Margin definition (Defined, Diffuse, N/A);
- Lesion density (Hypodense, Hyperdense, N/A);
- Benignancy (Benign, Rather Benign, Malignant, Rather Malignant, N/A);
- Lesion type (Cyst, HCC, Hemangioma, Metastasis, N/A).

While benignancy, lesion type and density annotations were used as class labels to train the RFs associated with each search context, the other annotated high–level features were tested as additional semantic input during the training phase of the models. After the models have been trained, at the retrieval stage, both the low–level features computed for the given lesion ROI and the high–level feature(s) are provided as input to the RF. Figure 2 in the next section shows a simple example illustrating the system workflow.

4   Framework Application to Similar Hepatic Lesion Retrieval

We apply the proposed retrieval framework to the search of similar hepatic lesions. The scheme shown in Figure 2 illustrates the search for hepatic lesions that have comparable benignancy given that the lesion specified in the ROI is known to have a defined margin.

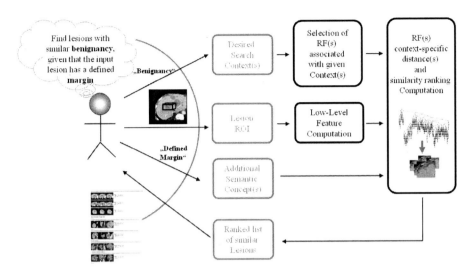

Fig. 2. Overview of the framework with an example application to the search of hepatic lesions with comparable benignancy (context), given that the lesion specified in the ROI has a defined margin (semantic concept or high–level feature).

In Figure 2, the input sample is represented by both a set of low–level features computed from the given ROI and by the given additional high–level concepts (in the example "Defined Margin"). The input sample is then run through the RF associated to (and trained for) the desired context ("Benignancy"). The database samples that occur in the same terminal nodes as the input sample will be noted and later ranked according to the frequency of co–occurrence with the input sample.

For hepatic lesion retrieval, three search contexts have been tested: lesion density, benignancy and type. For each of them, an associated Random Forest has been trained with the following annotated lesions from 523 3D CT scans corresponding to 244 patients:

Density (Hypodense, Hyperdense): 1948 hypodense and 358 hyperdense.
Benignancy (Benign, Malignant): 761 malignant and 93 benign.
Type (Cyst, Metastasis): 1103 metastases and 98 cysts.

Our experimental setting is flexible enough to accommodate any combination of high–level and low–level features, and has been tested in the 3 aforementioned contexts in a *Leave–One–Patient–Out* (LOPO) evaluation. We assess the retrieval results in the next section.

5   Empirical Study

We evaluate the retrieval performance of our framework in the three specified contexts, in terms of discrimination accuracy of the RF–based distance learnt.

We use LOPO ROC AUC (*Area Under the Curve*) values to evaluate the ultimate performance. The intrinsic RF similarity is compared with the canonical Euclidean distance. Both for the intrinsic RF similarity and the Euclidean metric, predictive performance is measured via k-nearest neighbor classification, with $k=7$ and with weighting the votes of the neighbors proportional to the RF similarity or inversely proportional to the Euclidean distance. This parameter setting has been shown to always lead to competitive results. For the Euclidean distance all feature values were also scaled to lie in the same range [0,1].

For each of the 3 search criteria and each representation of the input sample (ROI), Table 1 shows the ROC AUC measure of the retrieval results with the RF similarity, and Table 2 shows similar results for the Euclidean metric. The ROC curves pertaining to RFs trained in the three different contexts are shown in Figure 3. Table 3 shows semantic feature importance, i.e., the average appearence of a semantic feature given as input in a RF tree. Figure 4 illustrates the top 5 ranked results for 2 different input lesions.

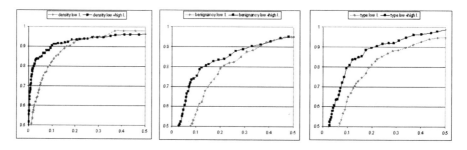

Fig. 3. ROCs for the training of the RFs for each context. From left to right, Density, Benignancy and Type context ROCs of the corresponding trained classifiers. In each case, the grey line shows the curve with the low–level features only, while the black line shows the curve when additional high–level features are included as input.

An additional high–level feature was tested in all three contexts: the location of the lesion (i.e. the segment number within the liver) was given as additional input to the framework. However, in all 3 contexts the feature had little or no influence on the results, which seems to suggest that the location of a lesion has little influence on its benignancy, density or type (a reasonable assumption).

Table 1 shows the results obtained using the intrinsic RF similarity. It is interesting to observe that the low–level features alone produce very reasonable retrieval results in the lesion density context, which is an inherently "low–level" task. High–level features seem indeed to have less impact in this context. However, when higher level clinical contexts such as benignancy or lesion type are considered, the positive impact of the additional high–level features given as input becomes more prominent, and results in a better accuracy and more relevant retrieval results. This benefit is particularly visible in the case of sub–centimeter lesions, which are often very hard to characterize even for the expert eye. In this

Table 1. Quantitative evaluation of retrieval results in the three contexts: Lesion Density, Benignancy and Type using, in each case, low–level features alone or a combination of low– and high–level features. The table shows the results for all lesion sizes combined, and sub–centimeter lesions are studied in the two rightmost columns.

	All Lesions		Lesions$\leq 1000mm^3$	
context	low l.	(low+high) l.	low l.	(low+high) l.
Density	0.94	0.933	0.959	0.955
Benignancy	0.854	0.891	0.732	0.855
Type	0.872	0.906	0.755	0.838
Average	0.889	0.91	0.828	0.8827

Table 2. Retrieval results with the Euclidean distance metric. Table has the same structure as Table 1.

	All Lesions		Lesions$\leq 1000mm^3$	
context	low l.	(low+high) l.	low l.	(low+high) l.
Density	0.55	0.543	0.525	0.542
Benignancy	0.72	0.713	0.655	0.655
Type	0.675	0.650	0.625	0.618
Average	0.648	0.635	0.601	0.605

case, the addition of one or more high–level features has a significant influence on the accuracy and meaningfulness of the results. The difference in performance in these cases (low–level features versus low- plus high-level features in the benignancy and lesion type contexts) is always significant according to the McNemar's test on difference between two proportions conducted for the specificity of interest (0.9). In a clinical setting where small lesions are difficult to assess, the proposed framework has great potential to become a significant decision support tool.

In the same 3 contexts we have also evaluated the Euclidean distance metric. The obtained results, shown in Table 2, are clearly inferior to those produced by the intrinsic RF similarity (the difference is always significant). Moreover, the Euclidean distance is not able to benefit from the addition of the high–level features, as opposed to the RF similarity, and the performance is often even worse after the addition of high–level features. The performance is especially poor in the lesion density context.

Table 3. Importance of the input high–level concepts for retrieval

	High–level feature given as additional input						
Context	Contrast Agent Phase	Focality	Surrounding	Rim	Margin	Margin Definition	Average
Density	1.358	1.393	1.443	1.081	1.052	0.964	1.215
Benignancy	1.801	1.641	1.771	1.091	2.293	1.89	1.749
Type	1.501	1.371	1.559	0.936	1.926	1.641	1.489
Average	1.553	1.468	1.591	1.0359	1.757	1.50	

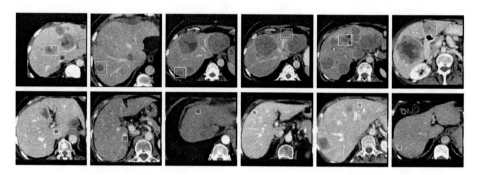

Fig. 4. Results of the search for similar lesions in the benignancy similarity context. Top row: A malignant lesion used as input (leftmost image), and the top 5 retrieved lesions, all of which are malignant as well, even though their appearances are not always comparable. Bottom row: Input benign lesion (leftmost image) and top 5 retrieval results, 4 of which are benign as well.

The time required for retrieval, from low–level feature computation to running the instance through the appropriate RF to similarity calculations and lesion sorting to the output of the ranked list of lesions, is less than a second on a standard PC. Although we consider three isolated user–specified search contexts in this study, a certain number of them can be easily combined when needed, for instance in order to address a query "Find lesions that are likely to be of same lesion type and have comparable density".

6   Conclusion and Future Work

We have presented a fast, flexible framework for semantic CBIR based on discriminative distance learning. The speed of retrieval, promising retrieval performance especially for difficult, small lesions, and the flexibility to define relevance criteria combinations with a considerable positive impact on the quality of the results open the door for higher level searches that yield results that are meaningful to the the user in any given context. The addition of semantic features as input for search consistently improves the retrieval results, with a particularly positive impact on small lesions, making results for all lesion sizes comparable in terms of quality and meaningfulness to the user.

We believe the framework presented in this paper has a large potential, as well as room for improvement. It would be interesting to allow not only the manual input of semantic features by the user, but also their automatic prediction, before they are used for retrieval. Using RF for regression tasks as well as multiclass–classification would allow the inclusion, for instance, of more lesion types to the similarity search. The presentation of results in the *html* report can also be complemented with more detailed statistics on the quality of the search results, which would allow the user to use the tool in a more informed way when making decisions.

Benefits of this framework include inconsistency checks in radiology reports (i.e. as a second opinion with respect to the reported diagnosis, benignancy, or any other lesion characteristics) and abnormality or relevance pointers highlighting critical lesions. In the future, the system will be combined with automatic lesion detection as a preprocessing step. This allows presenting the detected lesions to the radiologist in the form of a ranked list upon opening the case, prioritizing, for instance, the most likely malignant lesions. This will render diagnostic reading even more efficient by speeding up the expertŠs review of the ever increasing amount of image data.

References

1. Lencioni, R., Cioni, D., Bartolozzi, C., Baert, A.L.: Focal Liver Lesions: Detection, Characterization, Ablation. Springer, Heidelberg (2005)
2. Akgül, C.B., Rubin, D.L., Napel, S., Beaulieu, C.F., Greenspan, H., Acar, B.: Content–based image retrieval: current status and future directions. Journal of Digital Imaging (2010)
3. Napel, S.A., Beaulieu, C.F., Rodriguez, C., Cui, J., Xu, J., Gupta, A., Korenblum, D., Greenspan, H., Ma, Y., Rubin, D.L.: Automated retrieval of CT images of liver lesions on the basis of image similarity: Method and preliminary results. Radiology 256(1) (2010)
4. Spencer, J.A.: Indeterminate lesions in cancer imaging. Clinical Radiology 63, 843–852 (2008)
5. Smeulders, A.W.M., Worring, M., Santini, S., Gupta, A., Jain, R.: Content–based image retrieval at the end of the early years. IEEE Transaction on Pattern Analysis and Machine Intelligence 22(12) (2000)
6. Segal, E., Sirlin, C.B., Ooi, C., Adler, A.S., Gollub, J., Chen, X., Chan, B.K., Matchuk, G.R., Barry, C.T., Chang, H.Y., Kuo, M.D.: Decoding global gene expression programs in liver cancer by noninvasive imaging. Nature Biotechnology (2007)
7. Seifert, S., Thoma, M., Stegmaier, F., Hammon, M., Kramer, M., Huber, M., Kriegel, H.-P., Cavallaro, A., Comaniciu, D.: Combined semantic and similarity search in medical image databases 7967, 7967-2 (2011)
8. Pękalska, E.z., Harol, A., Duin, R.P.W., Spillmann, B., Bunke, H.: Non-Euclidean or Non-Metric Measures Can be Informative. In: Yeung, D.-Y., Kwok, J.T., Fred, A., Roli, F., de Ridder, D. (eds.) SSPR 2006 and SPR 2006. LNCS, vol. 4109, pp. 871–880. Springer, Heidelberg (2006)
9. Tsymbal, A., Huber, M., Zhou, S.K.: Learning discriminative distance functions for case retrieval and decision support. Transactions on CBR 3(1), 1–16 (2010)
10. Shi, T., Seligson, D., Belldegrun, A.S., Palotie, A., Horvath, S.: Tumor classification by tissue microarray profiling: random forest clustering applied to renal cell carcinoma. Mod Pathol. 18(4), 547–557 (2005)
11. Shi, T., Horvath, S.: Unsupervised learning with random forest predictors. Computational and Graphical Statistics 15(1), 118–138 (2006)
12. Hudak, A.T., Crookston, N.L., Evans, J.S., Hall, D.E., Falkowski, M.J.: Nearest neighbour imputation of species-level, plot-scale forest structure attributes from lidar data. Remote Sensing of Environment 112(5), 2232–2245 (2008)

13. Qi, Y., Klein-Seetharaman, J., Bar-Joseph, Z.: Random forest similarity for protein–protein interaction prediction from multiple sources. In: Prooceedings of Pacific Symposium on Biocomputing (2005)

14. Vitanovski, D., Tsymbal, A., Ionasec, R., Georgescu, B., Zhou, S.K., Comaniciu, D.: Learning distance function for regression-based 4d pulmonary trunk model reconstruction estimated from sparse MRI data. In: Proc. SPIE Medical Imaging (2011)

15. Breiman, L.: Random forests. Machine Learning, 5–32 (2001)

16. Saffari, A., Leistner, C., Santner, J., Godec, M., Bischof, H.: Online random forests. In: 3rd IEEE ICCV Workshop on Online Computer Vision (2009)

17. Oza, N., Russell, S.: Experimental comparisons of online and batch versions of bagging and boosting, pp. 359–364 (2001)

18. Pfahringer, B., Holmes, G., Kirkby, R.: New options for Hoeffding trees. In: Australian Conference on AI (2007)

19. Hu, M.K.: Visual pattern recognition by moment invariants. IRE Trans. Inform. Theory 8 (1962)

20. Pejnovic, P., Buturovic, L., Stojiljkovic, Z.: Object recognition by invariants. In: Proceedings of Int. Conf. on Pattern Recognition (1992)

Computer–Aided Diagnosis of Pigmented Skin Dermoscopic Images

Asad Safi[1,*], Maximilian Baust[1], Olivier Pauly[1], Victor Castaneda[1],
Tobias Lasser[1], Diana Mateus[1], Nassir Navab[1],
Rüdliger Hein[2], and Mahzad Ziai[2]

[1] Chair for Computer Aided Medical Procedures (CAMP)
Fakultät für Informatik, Technische Universität München, Germany
[2] Klinik und Poliklinik für Dermatologie und Allergologie
am Biederstein München, Technische Universität München, Germany
http://campar.in.tum.de/WebHome

Abstract. Diagnosis of benign and malign skin lesions is currently mostly relying on visual assessment and frequent biopsies performed by dermatologists. As the timely and correct diagnosis of these skin lesions is one of the most important factors in the therapeutic outcome, leveraging new technologies to assist the dermatologist seems natural. In this paper we propose a machine learning approach to classify melanocytic lesions into malignant and benign from dermoscopic images. The dermoscopic image database is composed of 4240 benign lesions and 232 malignant melanoma. For segmentation we are using multiphase soft segmentation with total variation and H^1 regularization. Then, each lesion is characterized by a feature vector that contains shape, color and texture information, as well as local and global parameters that try to reflect structures used in medical diagnosis. The learning and classification stage is performed using SVM with polynomial kernels. The classification delivered accuracy of 98.57% with a true positive rate of 0.991% and a false positive rate of 0.019%.

Keywords: Machine Learning, Classification, Supervised Learning, Melanoma, Computer–Aided Diagnosis.

1 Introduction

Skin cancer is among the most frequent types of cancer and one of the most malignant tumors. The incidence of melanoma in the general population is increasing worldwide [1]. Its incidence has increased faster than that of almost all other cancers, and the annual rates have increased on the order of 3% to 7% in the fair–skinned population in recent decades [1]. Currently, between 2 and 3 million non–melanoma skin cancers and 132.000 melanoma skin cancers

---

* Thanks to Deutscher Akademischer Austausch Dienst (DAAD), Higher Education Commission of Pakistan (HEC), Technische Universität München Graduate School (TUM-GS) and Klinikum rechts der Isar for the support in this research

H. Müller et al. (Eds.): MCBR-CDS 2011, LNCS 7075, pp. 105–115, 2012.

occur globally each year. One in every three cancers diagnosed is a skin cancer, and according to the Skin Cancer Foundation Statistics, one in every five Americans will develop skin cancer during their lifetime [2]. Because advanced cutaneous melanoma is still incurable, early detection, by means of accurate screening, is an important step toward mortality reduction. The differentiation of early melanoma from other pigmented skin lesions (e.g. benign neoplasms that simulate melanoma) is not trivial even for experienced dermatologists; in several cases, primary care physicians seem to underestimate melanoma in its early stage [3] which attracted the interest of many researchers, who have developed systems for automated detection of malignancies in skin lesions.

Dermoscopy consists in visual examination of skin lesion that are optically enlarged and illuminated by halogen light which is a non–invasive in vivo technique that assists the clinician in melanoma detection in its early stage [4,5]. This permits to visualize new morphologic features and in most cases facilitates early diagnosis. However, evaluation of the many morphologic characteristics is often extremely complex and subjective [6].

The Second Consensus Meeting on Dermoscopy was held in 2000 and its main conclusions were that four algorithms: pattern analysis, ABCDE rule, Menzies scoring method and the 7–point check list are good ways of evaluating skin lesions using dermoscopy. All four methods share some common concepts and allow for selection of specific features, which can be done with the aid of computers. The ABCDE rule specifies a list of visual features associated to malignant lesions (Asymmetry, Border unevenness, Color deviation, Dermoscopic structures and Elevation), from which a score is computed [7]. This methodology provided clinicians with a useful quantitative basis, but it did not prove efficient enough for clinically doubtful lesions (CDL). The main reason for this is the difficulty in visually characterizing the lesion's features. Setting an adequate decision threshold for the score is also a difficult problem; by now it has been fixed based on several years of clinical experience. Many authors claim that these thresholds may lead to high rates of false diagnoses [8]. The second most practiced algorithm for melanocytic lesion diagnosis is 7–points checklist [9]. This algorithm consists of analyzing the presence of the seven most important color or geometric structures that characterize malignant melanoma (blue whitish veil, atypical pigment network, irregular streaks, etc.).

Collaboration of dermatologists, computer scientists and image processing specialists has led to significant automation of analysis of dermoscopic images and improvement in their classification [10,11,12]. The computerized analysis of dermoscopic images can be an extremely useful tool to measure and detect sets of features from which dermatologists derive their diagnosis. It can also be helpful for primary screening campaigns, increasing the possibility of early diagnosis of melanoma. At present there is no commercial software used in clinical practice. Our conclusive aim is to develop software for the identification of early–stage melanomas, based on images obtained by digital dermoscopy. This would enable supervised classification of melanocytic lesions. The result of such classification procedure will separate the screened lesions into two groups. The first group corresponds to lesions that were classified with low confidence level which requires subsequent inspection by an experienced dermatologist for the final decision, while the second one corresponds to those lesions for which the confidence level is high and thus there is no need for examination by a dermatologist.

The paper is formulated as follows. In Section 2 we present a brief overview of previous related work. In Section 3 we describe the composition of our database of dermoscopic images, and in Section 4 we present our approach to melanocytic lesions classification. Results and performance are presented and discussed in Section 5. We conclude in Section 6.

2 Computerized Diagnosis of Dermoscopic Images: State of the Art

Computer aided image diagnosis for skin lesions is a comparatively new research field. While the first related work in the medical literature backdates to 1987, the contribution was limited since by that time computer vision and machine learning were both developing fields [13]. One of the first compelling contributions from the image processing community was reported from H. Ganster et al. [14]. In this paper, the researchers proposed a classical machine learning approach for dermoscopic image classification. The first stage is automatic, color–based lesion segmentation. Then, more than hundred features are extracted from the image (gradient distribution in the neighborhood of the lesion boundary, shape and color). Feature selection was obtained using sequential forward and sequential backward floating selection. Classification is performed using a 24–NN classifier and delivered a sensitivity of 77% with a specificity of 84%.

Up to our knowledge the best results in semi–automated melanocytic lesion classification where obtained by G. Capdehourat et al. [15]. The image database is composed of 433 benign lesions and 80 malignant melanoma. The learning and classification stage is performed using AdaBoost.M1 with C4.5 decision trees. For the automatically segmented database, classification delivered a false positive rate of 8.75% for a sensitivity of 95%. The same classification procedure applied to manually segmented images by an experienced dermatologist yielded a false positive rate of 4.62% for a sensitivity of 95%.

A complete summary of the results obtained by key studies from 2001 onwards are presented by Celebi et al. [16], along with their database sizes. As in [14], the proposed approach is a classic machine learning methodology. After an Otsu–based image segmentation, a set of global features are computed (area, aspect ratio, asymmetry and compactness). Local color and texture features are computed after dividing the lesion in three regions: inner region, inner border (an inner band delimited by the lesion boundary) and outer border (an outer band delimited by the lesion boundary). Feature selection is performed using Relieff [17] and CFS algorithms [18]. Finally, the feature vectors are classified into malignant and benign using SVM with model selection [19]. Performance evaluation gave a specificity of 92.34% and a sensitivity of 93.33%. Our contribution in this regards is a complete characterization of a skin lesions into a feature vector that contains shape, color and texture information, as well as local and global parameters that try to reflect structures used in medical diagnosis by dermatologists. The learning and classification stage is performed using SVM with polynomial kernels. The classification delivered accuracy of 98.57% with a true positive rate of 0.991% and a false positive rate of 0.019%.

3   Database

Our database is composed of 4472 images of melanocytic lesions: 4240 are benign lesions and 232 are malignant melanoma labeled by four dermatologists. It is important to note that in general these kind of lesions are the benign lesions that are visually the most alike to malignant melanoma; many of them are clinically doubtful even for experienced dermatologists. This data set is based on the existence of dermatoscopic and histopathologic studies, which were used as ground truth for the classification procedure.

4   Classification of Dermoscopic Images: Proposed Approach

Our approach follows a typical machine learning methodology. In the first stage, we tackle automatic segmentation to isolate the lesion's area from the normal skin. The second stage consists of feature extraction from the image for further lesion classification into malignant or benign. Features are inspired by the same elements that dermatologists use for lesion diagnosis ABCDE rule. Once the lesion's features have been extracted, labeled lesions are used to train a meta–classifier obtained using dataset balancing. Classification errors are obtained by means of cross validation. In this section we give details of each of these stages.

4.1   Segmentation

In order to segment the given image data we adapt the method described by Li et al. in [20]. Let $\Omega \subset \mathbb{R}^2$ denote the image domain. Then we define two soft–labeling functions $u_{1,2} : \Omega \to [0,1]$ which can be used to define three soft membership functions

$$M_1 = u_1 u_2, \quad M_2 = u_1(1 - u_2), \quad M_3 = 1 - u_1. \tag{1}$$

These membership functions provide a soft partitioning of the image domain, because $M_1(x) + M_2(x) + M_3(x) = 1$ holds for all $x \in \Omega$, and allow us to segment the image domain into three areas indicating healthy skin, bright parts of the melanoma, and dark parts of the melanoma. An example is shown in Figure 1.

The described partitioning of the image domain is obtained by minimizing the following convex energy

$$E = \frac{1}{2} \int_\Omega |\nabla u_1|^2 + |\nabla u_2|^2 \, dx + \lambda \sum_{k=1}^3 \int_\Omega d_k M_k \, dx, \tag{2}$$

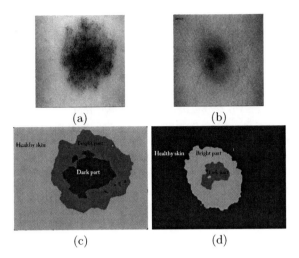

Fig. 1. Image (a),(b) Malignant melanoma and image (c),(d) segmented image in three areas

where

$$d_k = |a(x) - \bar{a}_k|^2 + |b(x) - \bar{b}_k|^2 . \tag{3}$$

Here $a, b : \Omega \to \mathbb{R}^3$ are the two color channels of the CIE Lab color space, while $\bar{a}_k, \bar{b}_k$ are the corresponding mean values:

$$\bar{a}_k = \frac{\int_\Omega M_k(x) a(x) \, dx}{\int_\Omega M_k(x) \, dx}, \quad \bar{b}_k = \frac{\int_\Omega M_k(x) b(x) \, dx}{\int_\Omega M_k(x) \, dx}. \tag{4}$$

The advantage of using the channels a and b of the CIE Lab color space is that these color channels only contain color and no luminance information making the segmentation more robust with respect to inhomogeneous lighting conditions. For all experiments we chose $\lambda = 2$. Please note that using an approach which minimizes a convex energy allows for a fully automatic segmentation of the data.

4.2 Feature Extraction

The feature extraction is the key point of the classification and has to be adequate in order to obtain a good system detection rate. We selected a group of features which attempts to represent the characteristics observed by the Physician. Each feature is following the idea of the ABCDE rules of skin cancer, which are:

- **A (Asymmetry):** Usually skin cancer moles are asymmetric instead of the normal moles, which are symmetric.
- **B (Border):** Usually the melanocytic lesions have blurry and/or jagged edges.
- **C (Color):** The melanocytic lesion has different colors inside the mole.

- **D (Diameter):** The lesions does not exceed a diameter of a pencil eraser (6 mm), otherwise it is suspicious.
- **E (Elevation):** When the mole is elevated from the normal skin it is suspicious.

Based on this technique, we created a set of features trying to characterize them via computer vision techniques. The list of features selected is as follows: geometric, color, texture and shape properties. The properties obtained by the feature extractor are totally based on the segmentation step and the features have to be independent of the image (size, orientation, etc.) in order to be robust with regard to the image acquisition. This feature property is very important because the physician can take the picture of the lesions in different ways, and lesions can have different sizes, too.

Geometric Properties: From segmentation of the lesions, we obtain a binary image which represents the segmented blobs. Using this binary image, we get the bounding box and we fit an ellipse which has the same second inertia moment of area. Smaller blobs are erased from the binary image. Usually the biggest blob of the image is the segmented lesion and the sparse small moles are only segmentation noise. The bounding box is our metric for the standardization of the lesions. Using the bounding box and the fitted ellipse we reorient the lesions to the biggest ellipse axis and we resize the image to a standard size. The features used to represent the geometric properties are as follows:

- **Relative Area:** Area of segmented mole with respect to the bounding box area. This area represents the size of the mole.
- **Relative Filled area:** Area of the segmented mole with the internal holes filled w.r.t. the bounding box area. It represents how many internal areas of the mole were wrongly segmented.
- **Relative Centroid:** The centroid of the fitted ellipse w.r.t. the bounding box, indicating the distribution of the mole in the bounding box.
- **Eccentricity:** The fitted ellipse eccentricity which represents how circular the mole is.
- **Solidity:** The relation between the convex area and the blob area, representing how irregular the border of the mole is.

We use the fitted ellipse and bounding box to **pre–process** the mole in order to create standard size and orientation to make the classification more robust. The orientation of one mole always will be the same because we apply a reorientation based not only on the orientation of the ellipse, but also on the largest distance of the blob border with regard to the centroid. These properties allow us to reorientate the same mole with different angles to the same orientation as shown in Figure 2. The bounding box is resized to a square using the largest side as the value of the square which is cropped and resized to a standard value of 100×100. This standard size allows to compare different moles with different sizes and orientation.

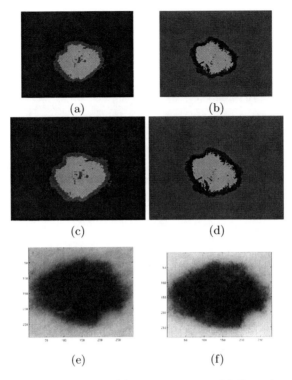

Fig. 2. Reorientation management: (a) First screening (b) Second screening (c) Segmentation of image a (d) Segmentation of image b (e) Reorientation of cropped image a (f) Reorientation of cropped image b

Color Properties: The mole color is very important in the classification because it encodes the variety of colors in the mole. When the mole has more colors, the mole has more chance to be malignant. The colors are coded in a color histogram representing all the colors observed in the mole. The histogram is compacted in groups of values named **bins**. The bins allow us to reduce the number of $256^3 \approx 16M$ entries of a sparse histogram to a reasonably small dense histogram. This reduction has the advantage of encapsulating different ranges of colors in only one histogram value and being more robust on lighting changes, but with the disadvantage of losing color precision. The selected number is 8 generating $8^3 = 512$ possible values in the histogram. The color histogram is created using only the pixels of the segmented mole, excluding the skin pixels. The histogram is normalized with the total number of pixels used to create the histogram. In this way, we can compare histograms created from different sized moles.

Texture Properties: This feature describes the differences between the colors of the mole allowing to characterize the discontinuity in the mole colors, which is a tool used by physicians to recognize if a mole is malignant or not. To represent the texture, we use LBP (local binary pattern) of the image which creates a code of the color variability in the neighborhood of each pixel.

Shape Properties: This feature represents the shape properties of the mole giving us how elliptic is the mole, circular or irregular, which is a very important feature in the classification of a mole. This feature is represented using histogram of oriented gradients (HOG), which counts the occurrences of gradients in portions of the image, coding the variability of the gradient in the image. This feature represents not only the shape of the moles, but also the mole uniformity given the internal shape when the color changes, which is detected by the gradient.

For each lesion, 8 features are extracted with vectors size 1682.

4.3   Classification

The goal of this stage is to classify the feature vectors in two classes: malignant and benign. A classification technique that proved very successful in our experiments are support vector machines (SVM, [21]). SVM was selected as the method of choice as it allows to linearly classify data in a high–dimensional feature space that is non–linearly related to the input space via the use of specific polynomial kernels. To reduce the dimensions of the input feature set $x_i \in R^{1682}$, $i = 1, ..., n$, where n denotes the number of measurements (in our case 4472), principal components analysis (PCA) is applied. The resulting feature vector of eigenvalues $(e_j^i)_{j=1,...,1682}$ is sorted descendingly by magnitude. Since the highest eigenvalues represent the most relevant components, a cut–off value C_{PCA} is chosen, such that the final input data y_i for the classification algorithm from measurement x_i $(i = 1, ...n)$ is

$$y_i = (e_j^i)_{j=1,...,C_{PCA}} \tag{5}$$

The cut–off value C_{PCA} is chosen empirically, to represent 95% of feature vector of 1682 dimensions, which reduced it to 434 dimensions.

The SVM classifier needs to be trained first before using it, thus we partition our input feature sets (y_i) $i = 1, .., n$, into two partitions, $T \subset \{1, ..., n\}$ the training set and $V \subset \{1, ..., n\}$ the testing (or validation) set with $T \cup V = \{1, ..., n\}$ and $T \cap V = \emptyset$. The training data set T is labeled manually into two classes using the ground truth, $l(y_i) = \pm 1$. Once the classifier is trained, a simple evaluation of the decision function $d(y_i) = \pm 1$ will yield the classification of any data y_i.

In detail, SVM is trying to separate the data $\phi(y_i)$ mapped by the selected kernel function ϕ by a hyperplane $w^T \phi(y_i) + b = 0$ with w the normal vector and b the translation. The decision function then is $d(y_i) = \text{sgn}(w^T \phi(y_i) + b)$. Maximizing the margin and introducing slack variables $\xi = (\xi_i)$ for non-separable data, we receive the primal optimization problem:

$$\min_{w,b,\xi} = \frac{1}{2} w^T w + C \sum_{i \in T} \xi_i \tag{6}$$

with constraints $l(y_i)(w^t\phi(y_i) + b) \geq 1 - \xi_i$, $\xi \geq 0$ for $i \in T$. C is a user–determined penalty parameter. Switching to the dual optimization problem allows for easier computation,

$$\min_{\alpha} = \frac{1}{2}\alpha^T Q\alpha - e^T\alpha \tag{7}$$

with constraints $0 \leq \alpha_i \leq C$ for $i \in T$, $\sum_{i \in T} y_i\alpha_i = 0$. The $\alpha = (\alpha_i)$ are the so–called support vectors, $e = [1, ...1]^T$ and Q is the positive semidefinite matrix formed by $Q_{jk} = l(y_j)l(y_k)K(y_j, y_k)$, and $K(y_j, y_k) = \phi(y_j)^T \phi(y_k)$ is the kernel function built from ϕ. Once this optimization problem is solved, we determine the hyperplane parameters w and b, w directly as $w = \sum_{i \in T} \alpha_i l(y_i)\phi(y_i)$ and b via one of the Karush-Kuhn-Tucker conditions as $b = -l(y_i)y_i^T w$, for those i with $0 < \alpha_i < C$. Thus the decision function of the trained SVM classifier ends up as

$$d(y_i) = \text{sgn}\big(w^T\phi(y_i) + b\big) = \text{sgn}\left(\sum_{j \in T} \alpha_i l(y_i)K(y_j, y_i) + b\right). \tag{8}$$

5 Results

Performance evaluation was conducted using a 10–fold cross–validation. The 10–fold cross–validation gives an approximation of the general classifier performance. We created 10 balanced data sets which were generated from the original unbalanced data set of 4240 benign and 232 malignant moles. The balanced data sets were generated by selecting randomly a similar number of benign and malign images (250) to obtain a more general and balanced training dataset. We assess the feature training and perform 10–fold cross–validation utilizing the 10 balanced datasets. The results of these data sets are shown in table 1

Table 1. Results of the 10 random balanced data sets, and for each dataset 10–fold cross–validation using a SVM classifier (Avg-Std 98.54414 ± 0.045317)

Variables	Test-1	Test-2	Test-3	Test-4	Test-5
Correctly Classified Instances	98.5772%	98.5743%	98.5765%	98.5614%	98.5167%
Incorrectly Classified Instances	1.4228%	1.4257%	1.4235%	1.4386%	1.4833%
True Positives Rate	0.991%	0.996%	0.993%	0.997%	0.995%
False Positives Rate	0.019%	0.023%	0.034%	0.025%	0.021%
	Test-6	Test-7	Test-8	Test-9	Test-10
Correctly Classified Instances	98.4982%	98.5765%	98.5965%	98.4624%	98.5017%
Incorrectly Classified Instances	1.5018%	1.4235%	1.4235%	1.5376%	1.4983%
True Positives Rate	0.981%	0.991%	0.983%	0.991%	0.996%
False Positives Rate	0.059%	0.033%	0.064%	0.020%	0.13%

The results show a very good performance in all the random data sets, allowing us to conclude that the selected feature vector of the moles gives meaningful information about the mole in the classification. The correctly classified instances value indicates a performance over 98% in all 10 tested cases and an error of less than 2%. If we observe only the malignant classification, which is the most important, the performance shows a true positives rate greater than 99%, meaning that the classifier recognizes as malignant 99% of the skin cancer moles. Therefore, the number of malignant moles which are not correctly classified is 1%. In addition, the false positives rate is smaller than 3%, showing that the misclassification of the benign images are only 3 in a total of 100 benign images. In the case of recognizing the malignant moles it is important to detect most of the malignant moles even if the false positive rate is not small, as it is less harmful for the physician to label the mole as suspicious even though it is not.

The Figure 3 shows the ROC response of our classifier and its consequential performance, having a curve near to the ideal case. The classifier has a high area under the curve being near to 0.99, where the maximum is 1. We believe that our feature vector is a good representation of the dermoscopy characteristics following the ABCDE rule used by the dermatologist in skin cancer diagnosis.

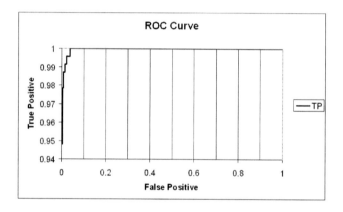

Fig. 3. Receiver Operating Characteristic response

6   Conclusions and Future Work

In this paper we proposed a methodology for computer–aided classification of dermoscopic images. The learning and classification stage is performed using SVM. According to our medical partners the results are satisfactory and for further research the system can be deployed in dermatology. Concerning our algorithm, to further improve its performance, methods to detect a larger number of geometric or texture–based structures, similar to those used in the 7–points checklist, should be developed. The next important step is sub–classification of malignant categories, which is ongoing research. A rigorous study of this topic, complemented with the comparison of the weights assigned to visual features in the ABCDE and other clinical diagnosis rules, may yield useful recommendations to dermatologist for their medical practice.

References

1. Marks, R.: Epidemiology of melanoma. Clin. Exp. Dermatol. 25, 459–463 (2000)
2. World Health Organization, Ultraviolet Radiation and the Intersun Programme (2007), http://www.who.int/uv/faq/skincancer/en/
3. Pariser, R.J., Pariser, D.M.: Primary care physicians errors in handling cutaneous disorders. J. Amer. Acad. Dermatol. 17, 239–245 (1987)
4. Carli, P., De Giorgi, V., Gianotti, B., et al.: Dermatoscopy and early diagnosis of melanoma. Arch Dermotal. 137, 1641–1644 (2001)
5. http://www.dermogenius.com, http://www.dermogenius.com
6. Rubegni, P., Burroni, M., Dell'eva, G., Andreassi, L.: Digital dermoscopy analysis for automated diagnosis of pigmented skin lesion. Clinics in Dermatology 20(3), 309–312 (2002)
7. Nachbar, F., Stolz, W., Merkle, T., Cognetta, A., Vogt, T., Landthaler, M., Bilek, P., Braun-Falco, O., Plewig, G.: The ABCD rule of dermatoscopy: high prospective value in the diagnosis of doubtful melanocytic skin lesions. Journal of the American Academy of Dermatology 30(4), 551–559 (1994)
8. Lorentzen, H., Weismann, K., Kenet, R., Secher, L., Larsen, F.: Comparison of dermatoscopic abcd rule and risk stratification in the diagnosis of malignant melanoma. Acta Derm Venereol 80(2), 122–126 (2000)
9. Johr, R.H.: Dermoscopy: alternative melanocytic algorithms - the abcd rule of dermatoscopy, menzies scoring method, and 7–point checklist. Clinics in Dermatology 20(3), 240–247 (2002)
10. Schmid-Saugeon, P., Guillod, J., Thiran, J.-P.: Towards a Computer–aided diagnosis System for Pigmented Skin Lesions, Comp. Med. Imag. Graphics, pp. 65–78 (2003)
11. Hall, P.N., Claridge, E., Smith, J.D.: Computer Screening for Early Detection of Melanoma: Is there a Future? British J. Dermatol. 132, 325–328 (1995)
12. Grzymala-Busse, P., Grzymala-Busse, J.W., Hippe, Z.S.: Melanoma prediction using data mining system LERS. pp. 615–620 (2001)
13. Cascinelli, N., Ferrario, M., Tonelli, T., Leo, E.: A possible new tool for clinical diagnosis of melanoma: The computer. Journal of the American Academy of Dermatology 16(2), 361–367 (1987)
14. Ganster, H., Pinz, A., Rhrer, R., Wildling, E., Binder, M., Kittler, H.: Automated melanoma recognition. IEEE Transactions on Medical Imaging vol 20, 233–239 (2001)
15. Capdehourat, G., Corez, A., Bazzano, A., Muse, P.: Pigmented Skin Lesions Classification Using Dermatoscopic Images (2009) ISBN: 978-3-642-10267-7
16. Celebi, M.E., Kingravi, H.A., Uddin, B., Iyatomi, H., Aslandogan, Y.A., Stoecker, W.V., Moss, R.H.: A methodological approach to the classification of dermoscopy images. Comput. Med. Imaging Graph 31(6), 362–373 (2007)
17. Robnik-Sikonja, M., Kononenko, I.: Theoretical and empirical analysis of relieff and rrelieff. Mach. Learn. 53(1-2), 23–69 (2003)
18. Hall, M.A.: Correlation–based feature selection for discrete and numeric class machine learning, pp. 359–366 (2000)
19. Schlkopf, B., Smola, A.J.: Learning with Kernels: Support Vector Machines, Regularization, Optimization and Beyond. The MIT Press, Cambridge (2001)
20. Li, F., Shen, C., Li, C.: Multiphase Soft Segmentation with Total Variation and H1 Regularization. Journal of Mathematical Imaging and Vision 37(2), 98–111 (2010)
21. Cristianini, N., Shawe-Taylor, J.: An Introduction to Support Vector Machines and Other Kernel-based Learning Methods. Cambridge University Press (2000) ISBN:0521780195

Texture Bags: Anomaly Retrieval in Medical Images Based on Local 3D-Texture Similarity

Andreas Burner[1,*], René Donner[1], Marius Mayerhoefer[2], Markus Holzer[1], Franz Kainberger[2], and Georg Langs[1,3]

[1] Computational Image Analysis and Radiology Lab, Department of Radiology, Medical University of Vienna, Austria
[2] Department of Radiology, Medical University of Vienna, Vienna, Austria
[3] CSAIL, Massachusetts Institute of Technology, Cambridge, MA, USA
andreas.burner@meduniwien.ac.at

Abstract. Providing efficient access to the huge amounts of existing medical imaging data is a highly relevant but challenging problem. In this paper, we present an effective method for content-based image retrieval (CBIR) of anomalies in medical imaging data, based on similarity of local 3D texture. During learning, a texture vocabulary is obtained from training data in an unsupervised fashion by extracting the dominant structure of texture descriptors. It is based on a 3D extension of the Local Binary Pattern operator (LBP), and captures texture properties via descriptor histograms of supervoxels, or *texture bags*. For retrieval, our method computes a texture histogram of a query region marked by a physician, and searches for similar bags via diffusion distance. The retrieval result is a ranked list of cases based on the occurrence of regions with similar local texture structure. Experiments show that the proposed local texture retrieval approach outperforms analogous global similarity measures.

Keywords: Content-based medical image retrieval (CBIR), feature extraction, texture analysis, unsupervised texture learning, Local Binary Pattern (LBP), localized features, high-resolution CT (HRCT), emphysema disease.

1 Introduction

In today's hospital routine, thousands of medical images are processed on a daily basis. Individual radiology departments can produce hundreds of gigabytes of high quality imaging data per day. This data is typically inspected once during diagnosis and then remains unused in the imaging repository of the hospital. Since this data together with the corresponding reports holds rich information,

* The research leading to these results has received funding from the European Union Seventh Framework Programme (FP7/2007-2013) under grant agreement n° 257528 (KHRESMOI) und from the Austrian Science Fund (FWF P22578-B19, PULMARCH).

H. Müller et al. (Eds.): MCBR-CDS 2011, LNCS 7075, pp. 116–127, 2012.

it is desirable to access it during diagnosis of new cases. However, currently this accessibility is limited, due to the complexity and amount of visual data. In this paper we propose a method for searching medical imaging data based on a query region marked by a physician.

State of the Art. Today's computer-aided detection (CAD) systems typically target one specific anatomic region. They consist of a classifier or scoring method that is trained to respond to a specific set of pathologies in a supervised fashion, using a priori knowledge regarding the characteristics of the disease and anatomical region. This approach does not scale well to cases that need large amounts of rich training data, or for which a priori knowledge regarding relevant features does not exist. In such scenarios we need methods that rely on at least partially unsupervised learning, to cope with the vast amount of data in an efficient and scalable manner.

Various methods have been introduced to analyze texture in digital images. A comparison of the two traditional approaches (statistical and structural) with the Local Binary Pattern (LBP) operator [18, 19] is discussed by Mäenpää et al. [14, 15]. The work concludes that statistical approaches, such as histograms of gray-level pixel values and gray-level co-occurrence (GLC) [10] work best with stochastic microtextures. In contrast, structural approaches such as Textons [12, 16], wavelet transforms and Gabor filters [17] compute weighted means of pixel values by applying filter banks etc. over a small neighborhood. They therefore work well with macrotextures, partially eliminating fine-grained information. The LBP operator combines both approaches. Therefore it works for stochastic microtextures as well as deterministic macrotextures.

Recent studies on lung tissue and soft tissue analysis show promising results [2–6, 20, 22, 23, 25]. However, the methods are based on supervised learning, and therefore are subject to the aforementioned limitations. Recently, unsupervised learning became a topic of interest in vision [7–9]. In contrast to supervised learning, unsupervised learning does not depend on annotated training sets, but learns domain specific structure from the data.

Aim. In this work we apply unsupervised learning to medical image retrieval. We learn a three-dimensional texture vocabulary that captures the imaged tissue properties specific to an anatomical domain and the associated anomalies. We do not use knowledge regarding the anomalies in the training cases, but instead learn the inherent structure in the visual data.

Typically soft tissue is characterized by a set of textures instead of homogeneous texture regions. Therefore, a description by a bag of descriptors instead of single descriptors is necessary. Our concept is analogous to the *visual word* paradigm of Sivic et al. [21] that is based on the idea that objects are represented by sets of typical patches of local appearance, called visual words. Histograms of the occurrence of these visual words are expected to be similar for objects within a class and distinct for objects of other classes. Our approach differs in two aspects from [21]: (1) instead of image patches we use three-dimensional Local Binary Pattern (LBP) texture descriptors to capture tissue characteristics,

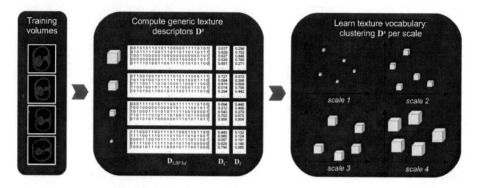

Fig. 1. Overview of the texture word learning pipeline: our descriptor $\mathbf{D}^s$ is a combination of a three-dimensional Local Binary Pattern (LBP) descriptor $\mathbf{D}_{LBP}$, a contrast measure $\mathbf{D}_C$, and an intensity measure $\mathbf{D}_I$. By applying k-means clustering for each scale s of $\mathbf{D}^s$ independently, we obtain generic texture words $\mathbf{W}_k^s$.

and (2) instead of using histograms to match entire images, we search for regions that have similar local appearance. This is necessary to cope with the variability encountered in medical imaging data, and the comparably subtle effects of disease. Based on the similarity of the texture structure of local regions, images are being ranked during retrieval.

The Scenario. The proposed method aims to support physicians during their diagnosis. A physician can mark a region of an image at hand and run a retrieval query on an image repository, for example the hospital's PACS system. Our method will return images with regions containing similar medical structures. This enables the physician to compare the current case to past cases and gain knowledge by comparing the diagnosis, treatment, and progress of the disease.

2   Method

The method consists of two phases. (1) During the learning phase (section 2.1) the algorithm computes descriptors and learns a three-dimensional texture vocabulary to capture the structure in the training data. (2) During the retrieval phase (section 2.2), the medical doctor marks a region of interest in the query image, the algorithm searches for similar regions in terms of texture, contrast, and intensity in the entire data set, and ranks images in the imaging repository accordingly.

2.1   Learning a Texture Vocabulary

This section describes how we train the texture vocabulary. Figure 1 shows the overview of the learning pipeline. The training set consists of medical images and a segmentation that marks the anatomical structure of interest.

Computation of Texture Features. The base feature extractor is a three-dimensional adaptation of the Local Binary Pattern operator (LBP) described

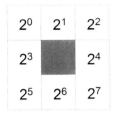 

Fig. 2. Local Binary Patterns (LBP) weights: a. Two-dimensional LBP by Ojala et al. [18, 19], b. Three-dimensional LBP used by our texture descriptor **D**.

by Ojala et al. [18, 19]. We chose to implement an LBP-based descriptor because of its property to respond very well to microscopic structure [14], and its computational simplicity and high performance. Furthermore, it is invariant to the gray-scale range, which makes it applicable for both, CT-, and MR imaging [11].

The original LBP operator computes the local structure at a given pixel i by comparing the values of its eight neighborhood pixels with the value of i, applying weights (figure 2a), and summing up the values. This computation yields a value from 0 to 255 for each pixel that describes the neighborhood-relative gray-scale structure. Various extensions of the original operator have been published in recent years[1], however, none of them addresses our focus.

A Three-dimensional, Multi-scale LBP Descriptor. As the local contrast is an important property in the medical domain, our method is based on LBP/C [19], which combines the base LBP operator $\mathbf{D}_{LBP}$, with a local contrast measure $\mathbf{D}_C$. In some medical imaging modalities, such as CT, intensities of local regions are an important decision instrument for physicians. Since the LBP operator is by definition gray-scale invariant, we furthermore supplement a local average intensity measure $\mathbf{D}_I$ of the 3x3x3 LBP cube to the feature vector if absolute intensity is relevant. We denote our LBP3d/CI descriptor **D**, defined as:

$$\mathbf{D} = [\,\mathbf{D}_{LBP3d}, c_c\mathbf{D}_C, c_i\mathbf{D}_I\,] \tag{1}$$

In total the descriptor **D** has 28 dimensions: 26 LBP bits, one contrast dimension, and one intensity dimension. The factors c_c and c_i determine the impact of the two measures, contrast and intensity. In practice, all descriptors are scaled to the range $[0, 1]$, c_c and c_i are chosen according to the imaging modality.

Note that extending the LBP operator from 2D to 3D increases the dimensionality of the feature space substantially. In 2D the 3x3 pixel grid results in a descriptor of 8 bit size (3^2-1) (figure 2a), whereas in 3D the 3x3x3 pixel grid results in a descriptor of 26 bit size (3^3-1) (figure 2b). The increased dimensionality causes a large increase of possible texture units, from 256 to $67.1*10^6$.

---

[1] An extensive list of LBP bibliography can be found at http://www.cse.oulu.fi/MVG/LBP_Bibliography (accessed June 2011).

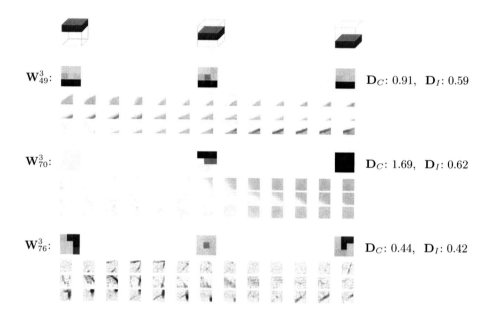

$\mathbf{W}_{49}^{3}$: $\mathbf{D}_C$: 0.91, $\mathbf{D}_I$: 0.59

$\mathbf{W}_{70}^{3}$: $\mathbf{D}_C$: 1.69, $\mathbf{D}_I$: 0.62

$\mathbf{W}_{76}^{3}$: $\mathbf{D}_C$: 0.44, $\mathbf{D}_I$: 0.42

Fig. 3. Three example 3D *texture words* $\mathbf{W}_{49}^{3}$, $\mathbf{W}_{70}^{3}$, and $\mathbf{W}_{76}^{3}$ trained in an unsupervised manner on lung tissue on scale 3. The slices of the cube from top to bottom are depicted from left to right. Next to the cube are the measures for contrast ($\mathbf{D}_C$) and intensity ($\mathbf{D}_I$). Below each *texture words* are three examples of lung tissue belonging to this word. Note the similar structure of the example-tissues for each *texture word*.

Learning a Vocabulary of Local Patterns: *Texture Words*. To obtain structure from the data and quantize this enormous feature space, we perform clustering on the feature descriptors $\mathbf{D}$ of a large set of voxels randomly sampled from the anatomical structure across the training set. After computing k-means, the centroids define the texture vocabulary for a specific anatomical region. We refer to the k clusters that formulate the texture vocabulary as *texture words* $\mathbf{W}_k$. Note, that the centroid's coefficients that correspond to the binary LBP features are now in the interval $[0, 1]$. We compute Euclidean distance as the metric for k-means clustering, as each bit of the LBP code represents one dimension with a binary value of either 0 or 1 and contrast and intensity are also in Euclidean space.

In the medical domain, the scale in which a texture appears is of high relevance. The importance of scale can be seen in the approach of André et al. [1]. To be able to perceive the granularity of texture, the method incorporates a multi-resolution approach by computing the descriptor $\mathbf{D}^s$ on various scales of the volumes. By performing clustering on the descriptors of each scale s independently, the learned vocabulary $\mathbf{W}_k^s$ is specific for each scale.

Analogous to Textons [12, 16], we represent each voxel with its closest texture word $\mathbf{W}_k^s$, i.e., with the index of the closest cluster center. Figure 3 shows three examples of cluster centers, or *texture words*.

Interpretation of Texture Words. Three texture words resulting from clustering the descriptors **D**, together with corresponding instances in the data are illustrated in figure 3. To evaluate the texture words, random examples of tissue patches have been visually inspected and compared. Note that the three examples describe very different characteristic local texture volumes of lung tissue - a result of **D** being a combination of $\mathbf{D}_{LBP3d}$, $\mathbf{D}_C$, and $\mathbf{D}_I$. The center voxel of the cube is gray, surrounded by darker or lighter voxels describing the relative structure to the center. $\mathbf{W}^3_{49}$ represents a strong vertical edge, whereas $\mathbf{W}^3_{70}$ represents a diagonal edge, and $\mathbf{W}^3_{76}$ represents a horizontal structure on the top of the cube and a diagonal structure at the bottom. The low value of c furthermore describes only little, but strong structures.

Properties of Texture Words and the Concept of Texture Bags. We describe a region **R** by the histogram $h(\mathbf{R})$ of *texture words* $\mathbf{W}_k$ it contains. This is analogous to the *bag of visual word* paradigm of Sivic et al. [21]. We call this k-bin histogram $h(\mathbf{R})$ a *texture bag*. Similar to [21] describing local patches, our histogram describes a region in terms of its textural structure. We normalize the texture histogram to make comparison possible without considering the size of a region.

A Distance between Texture Bags. Comparing texture histograms is not trivial since similarity among texture words can make bin to bin comparison ambiguous. To take the similarity of texture words into account, when computing the distance between texture bags, we use the diffusion distance algorithm described by Ling et al. [13]. It takes the relationship among words into account in form of a weight matrix. Based on our definition of texture words, we can compute cross-word weights c_{ij} as the dimension-wise distance between two words **W**:

$$c_{ij} = \sqrt{\mathbf{W}_i{}^2 - \mathbf{W}_j{}^2}, \tag{2}$$

and create the matrix $\mathbf{C} = [c_{ij}]$ by computing this distance for all word pairs. Diffusion distance has the added benefit of increasing the robustness of the method with regard to the choice of k during clustering.

The distance between two *texture bags* $\mathbf{R_i}$ and $\mathbf{R_j}$ is computed by

$$d_{ij} = d(\, h(\mathbf{R}_i), h(\mathbf{R}_j)\,) \tag{3}$$

2.2   Retrieval

For the scenario described in the introduction, the retrieval query consists of a medical image $\mathbf{I}_Q$ and a marked query region $\mathbf{R}_Q$. Our method aims to retrieve images with regions most similar to $\mathbf{R}_Q$. To compare the texture of areas, the corresponding texture bags are compared by the diffusion distance. Figure 4 shows an overview of the retrieval pipeline.

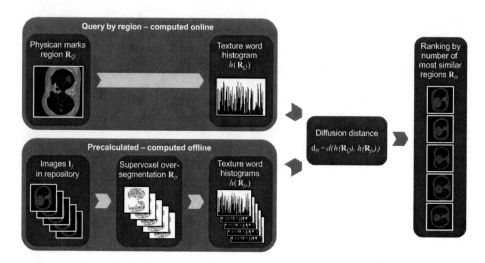

Fig. 4. Overview of the 3D-LBP/CI retrieval pipeline. Top left: the retrieval scenario, a physician marks a region in an image. Lower left: precomputation of the image set. Right side: comparison by diffusion distance [13] and ranking by the number of most similar regions.

Precomputation of the Images in the Imaging Repository. For the purpose of grouping similar areas and reducing the complexity of the three-dimensional image, we perform a precomputing step for each image $\mathbf{I}_j$ of the repository. This precomputing step fragments each image into several texture bags. We chose to use a supervoxel algorithm for this purpose and apply the method of Wildenauer et al. [24]. The result for a lung volume $\mathbf{I}_j$ is a three-dimensional oversegmentation $\mathbf{R}_{js}$ for the image j and the supervoxel index s, shown in figure 5. For each region $\mathbf{R}_{js}$, we precompute a *texture bag*, a histogram $h(\mathbf{R}_{js})$ of occurring texture words $\mathbf{W}_k$.

Fig. 5. Supervoxel algorithm applied on lung volumes. This figure depicts the overseg-mented regions $\mathbf{R}_{js}$ in 2D on the left, and in 3D on the right.

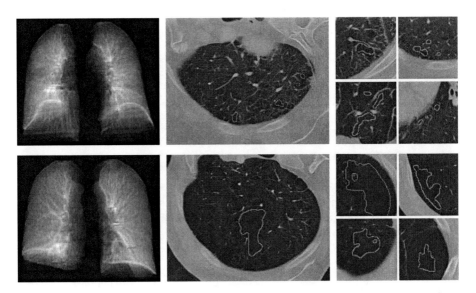

Fig. 6. Retrieval ranking result of two distinct emphysemas with different tissue patterns (top: centrilobular emphysema, bottom: panlobular emphysema). The region highlighted in red on the left side shows the query region $\mathbf{R}_Q$ marked during search by a physician. On the right side, the green regions depict the four most similar regions $\mathbf{R}_{js}$ retrieved by our method.

Computing Similarities to the Retrieval Query. To compare the marked regions $\mathbf{R}_Q$ of the query image $\mathbf{I}_Q$ to all regions $\mathbf{R}_{js}$, a normalized texture word histogram $h(\mathbf{R}_Q)$ is computed. The distance between the histogram of the query region $\mathbf{R}_Q$ and the regions $\mathbf{R}_{js}$ is computed by the diffusion distance.

$$\mathbf{d}_{js} = d(\, h(\mathbf{R}_Q), h(\mathbf{R}_{js})\,) \tag{4}$$

2.3 Ranking of the Image Set

The final step of the retrieval pipeline computes a ranking of all images $\mathbf{I}_j$ based on the number of "close" regions $\mathbf{R}_{js}$ to the query region $\mathbf{R}_Q$. For this ranking, the number of regions $\mathbf{R}_{\mathbf{I},s}$ are considered that are amongst the most similar regions $\mathbf{R}_{js}$ to $\mathbf{R}_Q$, in terms of diffusion distance. The threshold t of regions taken into account is dependent on the average region size $\mathbf{R}_{js}$, therefore dependent on the number of superpixels s per image $\mathbf{I}_j$.

Figure 6 shows the result of two retrieval queries. The first example is a query to retrieve patterns that are typical for centrilobular emphysema: round black spots, with typically less tissue structure than healthy lung tissue. The second example query retrieves patterns that are characteristic for panlobular emphysema: large areas with very little tissue lung structure. The results on the right side show regions with small distance to the query region $\mathbf{R}_Q$, i.e., regions where the texture histograms $h(\mathbf{R}_Q)$ and $h(\mathbf{R}_{js})$ are similar.

3   Data

The evaluation data contains 21 HRCT image series with a slice thickness of 3mm and an in-plane pixel spacing of 0.74mm. The cases of this study consist of 10 cases of lungs diagnosed as healthy and cases with two types of lung pathologies: emphysema (6 cases), lung metastasis (4 cases), and both (1 case). The diagnosis of each case was confirmed by two experienced radiologists of the contributing hospital.

In this paper we focus on the retrieval of lungs suffering emphysema. To simulate the retrieval scenario, a radiologist marked a query region in the query case. Query cases were emphysema cases. The system retrieves most similar regions in the remaining data set and ranks them accordingly to the distance (Eq. 4).

As a basis for our tests, the contributing radiologists manually marked query regions in each lung where they detected patterns of emphysema. These marked regions typically are small patches on three to five slices for each image series (see left images of figure 6).

The segmentation of the lungs are performed semi-automatically, by a threshold algorithm, applying simple morphologic functions, and a manual validation and correction step. Lung segmentation is not in the scope of this paper. Instead we focus on the characterization of anomalies within the anatomical structure.

4   Experiments

4.1   Set-Up and Evaluation

To achieve clinically applicable performance, the code for the three-dimensional, LBP-based descriptor $\mathbf{D}$ is implemented in C++. Therefore, the processing time for a volume of 512x512x150 voxels is less than one second on a quad core computer.

During our pilot experiments that are based on the lung anomaly retrieval scenario, we chose fixed parameters for all runs. We chose the number of clusters $k = 300$. Note that the diffusion distance results in some degree of robustness regarding increasing k. Our tests show that the performance of our method is to some extent dependent on the precomputed oversegmentation of the volumes. Therefore, the supervoxel algorithm is of importance: (1) where it computes the borders between regions, and (2) the number of supervoxels s per volume, which should be chosen dependent on the granularity of the anomaly to be retrieved. After initial experiments, we fixed $s = 5000$ for all cases. For the descriptor $\mathbf{D}$ we chose the weights c_c and c_i to be 10, and use scale 1 to 4 for the similarity calculation of regions. Furthermore, we set the ranking threshold to 800.

We performed a 7-fold cross validation to evaluate the retrieval performance of the algorithm. In each run, a *query region* $\mathbf{R}_Q \subset \mathbf{I}_Q$ was marked in one of the emphysema cases, and retrieval was performed on all other cases $\mathbf{I}_j, j \neq Q$. We validate the ranking by evaluating the ratio of cases with corresponding anomaly (emphysema) among the top ranked retrieved cases.

To validate the concept of *texture bag* ranking based on a local query region, the recognition rate of our method (table 2) was compared to a retrieval run based on the distance of *texture word* histograms of the entire query image $h(\mathbf{I}_Q)$ to texture word histograms of entire lung volumes of the data set $h(\mathbf{I}_j), j \neq Q$.

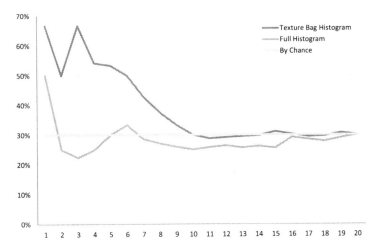

Fig. 7. Comparison of retrieval results: (yellow) ratio of randomly picked image series is 30%. Our method (blue) that is based on unsupervised learning outperforms full volume histogram retrieval (green).

4.2   Results

Table 1 shows anomaly retrieval results for global image similarity, table 2 demonstrates the retrieval results by the proposed method. The ratio of correctly top ranked cases (i.e., cases with the correct query anomaly) for randomly picked image series is 30%. The full volume histogram retrieval's performance is in the range of 24%-29%. The likely reason is that the anomaly regions that are relevant for retrieval often cover only a fraction of the overall volume. Therefore, the most frequent tissue types are typically not anomalies, and thus not helpful

Table 1. Full image ranking: the result of seven runs (r1 to r7): the table shows the number of correctly retrieved anomalies by full volume histogram retrieval of the top 3, 5, and 7 image series

ranking	r1	r2	r3	r4	r5	r6	r7	average
top 3	1	0	0	1	1	1	1	24%
top 5	2	0	1	2	2	2	1	29%
top 7	2	1	1	3	2	3	1	27%

Table 2. Texture bag ranking: the result of seven runs (r1 to r7): the table shows the number of correctly retrieved anomalies by our retrieval method of the top 3, 5, and 7 image series

ranking	r1	r2	r3	r4	r5	r6	r7	average
top 3	2	2	2	2	2	1	3	67%
top 5	2	3	4	3	2	2	3	54%
top 7	2	3	4	4	3	3	3	45%

during retrieval. The average recognition rate of correctly retrieved anomalies of all runs for the proposed method is 67% for the top rated three, 54% for the top rated five, and 45% for the top rated seven image series. This shows that the retrieval performance of our method outperforms retrieval by chance and full volume histogram retrieval significantly. Figure 7 plots the percentage of correctly retrieved anomalies against the number of image series returned.

5   Conclusion

In this paper we present a fast method for content-based image retrieval (CBIR) based on unsupervised learning of a vocabulary of local texture. We describe the concept of *texture bags*, an efficient way to compare regions based on their texture content. The algorithm is suitable for anomaly retrieval in large image repositories, such as PACS systems. On an initial data set, our experiments indicate that the proposed retrieval method outperforms global similarity measures if the aim is to retrieve cases with the same anomaly. Future work will focus on the evaluation on data sets with clinically realistic size and a wide variety of anomalies.

References

1. André, B., Vercauteren, T., Perchant, A., Buchner, A.M., Wallace, M.B., Ayache, N.: Endomicroscopic image retrieval and classification using invariant visual features. In: Proceedings of the Sixth IEEE International Conference on Symposium on Biomedical Imaging: From Nano to Macro, ISBI 2009, pp. 346–349. IEEE Press, Piscataway (2009)
2. Depeursinge, A., Iavindrasana, J., Hidki, A., Cohen, G., Geissbühler, A., Platon, A., Poletti, P., Müller, H.: Comparative performance analysis of state-of-the-art classification algorithms applied to lung tissue categorization. J. Digital Imaging 23(1), 18–30 (2010)
3. Depeursinge, A., Vargas, A., Gaillard, F., Platon, A., Geissbuhler, A., Poletti, P., Müller, H.: Content-based retrieval and analysis of HRCT images from patients with interstitial lung diseases: a comprehesive diagnostic aid framework. In: Computer Assited Radiology and Surgery (CARS) 2010 (June 2010)
4. Depeursinge, A., Vargas, A., Platon, A., Geissbuhler, A., Poletti, P.-A., Müller, H.: 3D Case–Based Retrieval for Interstitial Lung Diseases. In: Caputo, B., Müller, H., Syeda-Mahmood, T., Duncan, J.S., Wang, F., Kalpathy-Cramer, J. (eds.) MCBR-CDS 2009. LNCS, vol. 5853, pp. 39–48. Springer, Heidelberg (2010)
5. Depeursinge, A., Zrimec, T., Busayarat, S., Müller, H.: 3D lung image retrieval using localized features. In: Medical Imaging 2011: Computer-Aided Diagnosis, vol. 7963, p. 79632E. SPIE (February 2011)
6. Fehr, J., Burkhardt, H.: 3D rotational invariant local binary patterns. In: Proceedings of the 19th International Conference on Pattern Recognition (ICPR 2008), Tampa, Florida, USA, pp. 1–4 (2008)
7. Fergus, R., Perona, P., Zisserman, A.: Object class recognition by unsupervised scale-invariant learning. In: IEEE Computer Society Conference on Computer Vision and Pattern Recognition, vol. 2, p. 264 (2003)

8. Fritz, M., Schiele, B.: Towards Unsupervised Discovery of Visual Categories. In: Franke, K., Müller, K.-R., Nickolay, B., Schäfer, R. (eds.) DAGM 2006. LNCS, vol. 4174, pp. 232–241. Springer, Heidelberg (2006)

9. Grauman, K., Darrell, T.: Unsupervised learning of categories from sets of partially matching image features. In: Proceedings of the 2006 IEEE Computer Society Conference on Computer Vision and Pattern Recognition, vol. 1, pp. 19–25. IEEE Computer Society (2006)

10. Haralick, R.M., Shanmugam, K., Dinstein, I.: Textural features for image classification. IEEE Transactions on Systems, Man, and Cybernetics SMC-3, 610–621 (1973)

11. Hou, Z.: A Review on MR Image Intensity Inhomogeneity Correction. International Journal of Biomedical Imaging, 1–12 (2006)

12. Leung, T., Malik, J.: Recognizing surfaces using three-dimensional textons. In: Proceedings of the International Conference on Computer Vision, ICCV 1999, vol. 2, pp. 1010–1017. IEEE Computer Society, Washington, DC (1999)

13. Ling, H., Okada, K.: Diffusion distance for histogram comparison. In: Proceedings of the 2006 IEEE Computer Society Conference on Computer Vision and Pattern Recognition, vol. 1, pp. 246–253 (2006)

14. Mäenpää, T.: The Local Binary Pattern Approach To Texture Analysis Extensions And Applications (Academic Dissertation), p 20. University of Oulu (August 2003)

15. Mäenpää, T., Pietikäinen, M.: Texture Analysis With Local Binary Patterns, pp. 197–216. World Scientific Publishing Co. (January 2005)

16. Malik, J., Belongie, S., Leung, T., Shi, J.: Contour and texture analysis for image segmentation. Int. J. Comput. Vision 43, 7–27 (2001)

17. Manjunath, B.S., Ma, W.Y.: Texture features for browsing and retrieval of image data. IEEE Trans. Pattern Anal. Mach. Intell. 18, 837–842 (1996)

18. Ojala, T., Pietikainen, M., Harwood, D.: Performance evaluation of texture measures with classification based on Kullback discrimination of distributions, vol. 1, pp. 582–585. IEEE (1994)

19. Ojala, T., Pietikainen, M., Harwood, D.: A comparative study of texture measures with classification based on featured distributions. Pattern Recognition 29(1), 51–59 (1996)

20. van Rikxoort, E., Galperin-Aizenberg, M., Goldin, J., Kockelkorn, T., van Ginneken, B., Brown, M.: Multi-classifier semi-supervised classification of tuberculosis patterns on chest ct scans. In: The Third International Workshop on Pulmonary Image Analysis, pp. 41–48 (2010)

21. Sivic, J., Zisserman, A.: Video Google: A text retrieval approach to object matching in videos. In: Proceedings of the Ninth IEEE International Conference on Computer Vision, vol. 2, pp. 1470–1477. IEEE Computer Society (2003)

22. Sørensen, L., Shaker, S.B., de Bruijne, M.: Texture Classification in Lung CT using Local Binary Patterns. In: Metaxas, D., Axel, L., Fichtinger, G., Székely, G. (eds.) MICCAI 2008, Part I. LNCS, vol. 5241, pp. 934–941. Springer, Heidelberg (2008)

23. Tolouee, A., Abrishami-Moghaddam, H., Garnavi, R., Forouzanfar, M., Giti, M.: Texture analysis in lung HRCT images. Digital Image Computing: Techniques and Applications, 305–311 (2008)

24. Wildenauer, H., Mičušík, B., Vincze, M.: Efficient Texture Representation using Multi-Scale Regions. In: Yagi, Y., Kang, S.B., Kweon, I.S., Zha, H. (eds.) ACCV 2007, Part I. LNCS, vol. 4843, pp. 65–74. Springer, Heidelberg (2007)

25. Zavaletta, V.A., Bartholmai, B.J., Robb, R.A.: High resolution multidetector CT-aided tissue analysis and quantification of lung fibrosis. Academic Radiology 14(7), 772–787 (2007)

Evaluation of Fast 2D and 3D Medical Image Retrieval Approaches Based on Image Miniatures

René Donner[1,*], Sebastian Haas[1], Andreas Burner[1],
Markus Holzer[1], Horst Bischof[2], and Georg Langs[1,3]

[1] Computational Image Analysis and Radiology Lab, Department of Radiology,
Medical University of Vienna, Austria
[2] Institute for Computer Graphics and Vision,
Graz University of Technology, Austria,
[3] Computer Science and Artificial Intelligence Laboratory,
Massachusetts Institute of Technology, Cambridge, MA, USA
rene.donner@meduniwien.ac.at

Abstract. The present work evaluates four medical image retrieval approaches based on features derived from image miniatures. We argue that due to the restricted domain of medical image data, the standardized acquisition protocols and the absence of a potentially cluttered background a holistic image description is sufficient to capture high-level image similarities. We compare four different miniature 2D and 3D descriptors and corresponding metrics, in terms of their retrieval performance: (A) plain miniatures together with euclidean distances in a k Nearest Neighbor based retrieval backed by kD-trees; (B) correlations of rigidly aligned miniatures, initialized using the kD-tree; (C) distribution fields together with the l_1-norm; (D) SIFT-like histogram of gradients using the χ^2-distance. We evaluate the approaches on two data sets: the ImageClef 2009 benchmark of 2D radiographs with the aim to categorize the images and a large set of 3D-CTs representing a realistic sample in a hospital PACS with the objective to estimate the location of the query volume.

Keywords: medical image retrieval, 3D retrieval, image descriptors, anatomical region localization.

1 Introduction

In a typical clinical setting radiologists base their diagnosis on the image data of the patient at hand, drawing on their expertise (i.e. acquired, implicit image models) or reference textbooks. Typically the wealth of information present in

* The research leading to these results has received funding from the Austrian Sciences Fund (P 22578-B19, PULMARCH), the Austrian National Bank Anniversary Fund (13497, AORTAMOTION) and the EU-funded KHRESMOI project (FP7-ICT-2009-5/257528).

H. Müller et al. (Eds.): MCBR-CDS 2011, LNCS 7075, pp. 128–138, 2012.

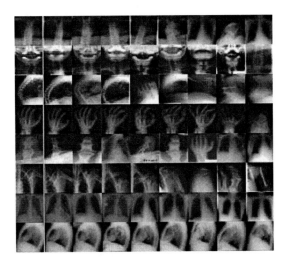

Fig. 1. The aim of content-based medical image retrieval: Given a query image (first column), find the most similar images from a given set, in this case the ImageCLEF 2009 data set

every hospital's PACS in the form of image data of prior patients is not exploited. The main reason is a time-consuming retrieval process that relies on textual/categorical information. The present work focuses on the first step required for a system that utilizes the existing image data. The goal is to efficiently retrieve a subset of overall visually highly similar images for a given query image. The resulting set of images can be either directly displayed or passed on to modality, region or pathology specific analysis stages. For example, this is relevant in the context of anatomical structure localization approaches such as [4], where separate models need to be employed for each anatomical region.

We argue that the characteristics of medical images, and radiological data in particular, warrant a closer look at holistic image representations as a means for this type of medical image retrieval. Medical imaging protocols are standardized, meaning the positioning of body parts in the image varies very little within a set of images of the same body region. A large proportion of the acquired images are from a few body regions and view points, and overall imaging characteristics (i.e. post-processing kernels in lung CTs) are similar for similar diagnostic questions.

1.1 Descriptor Formation in Medical Image Retrieval

There are two widely used approaches to obtain image descriptors that are comparable despite differences in image size and content. On the one hand we can construct a visual vocabulary of so-called visual words based on local descriptors around interest points, on the other hand we can use global image features such as color or gray level distributions or statistics over filter responses. In the context of the ImageCLEF benchmark, medGIFT[1] makes use of the latter approach, while also taking into account textual and image size cues.

Bags of Visual Words cluster the image descriptors from a large set of training images and use the cluster centers (the visual words) to derive one histogram (bag) for each image: all descriptors of one (training or test) image are attributed to the respectively closest cluster and the C bins of each histogram reflect these occurrences. Image similarities are then computed using a χ^2-metric on the normalized histograms. For an application of BVWs in the medical context see e. g. [1,5]. [1] also investigates the stability of the commonly used Differences of Gaussians (DOG) interest point detector for medical data, and shows that using a fixed grid of interest points can yield better results.

A third alternative are holistic approaches. The present work maps all images into a common space of 32×32 pixel miniatures (or 16^3 voxels for volumes) and performs subsequent computations therein. As opposed to standard BVWs and histograms over the entire image, the spacial relations of the image subparts are not lost, while the derived descriptors are considerably cheaper to compute than BVW representations (both during training and for each query).

1.2   Related Work

On the ImageCLEF 2009 Medical Annotation Task data set, *TAUbiomed* employ BVWs by using 9x9 pixel patches as descriptors. Classification is preformed by a support vector machine (SVM) using the resulting visual word histograms [2]. Dimitrovski et al. propose a hierarchical multi-label classification, using edge histogram descriptors as global and SIFT descriptors and Local Binary Patterns (LBP) as local features [3]. Tommasi et al. [11] also use an BVWs approach together with a χ^2-distance kernel SVM. Unay et al. also used LBP as features with hierarchical SVM [13]. medGIFT on the other hand represents an approach where image features are computed without a BVWs framework. Gabor filter responses on different scales are employed together with mean gray values of iteratively subdivided image areas. Feature selection is employed to reduce the potentially large feature space.

While some of the above approaches also incorporate features derived from small scale images of the data set, [7] directly used the idea of rigidly aligning thumbnails of query images. It was used to align each query image with selected prototype images for each category. [12] employ the same idea on massive data sets: Based on the finding that humans can recognize image scenes on images as small as 32 by 32 pixels, thumbnails of this size are used as sole descriptors of the images. The feasibility of this approach has to be seen in the context of the task and the number of data: Given the data set of 80 million images the space of training images is comparatively highly populated. While querying with a test image does not necessarily guarantee finding images of the same scene or object, but at least visually highly similar images, within which the user is likely to find images related to his or her query.

---

[1] MedGIFT Medical Image Retrieval Framework: http://medgift.hevs.ch/

2   Methods

In the following the construction of the four descriptor / metric pairs is detailed, miniatures with and without rigid alignment, distribution fields and SIFT-like histograms of gradients.

2.1   Image Miniatures as Descriptors

While the number of available training images in medical data sets is smaller than in [12], we argue that due to the constrained domain - only radiographs or CTs, only of the human body - the training space can also be considered to be densely populated and that a euclidean distance between image descriptors formed from miniature images provides a reasonable initialization for further optimization.

Given the set of training images $\mathcal{I} = \{\mathbf{I}_1, \ldots, \mathbf{I}_N\}$ each image is rescaled to size 32×32 to form the descriptors $\mathbf{D} = (\mathbf{d}_i, \ldots, \mathbf{d}_N)$. PCA is applied to $\mathbf{D}$, retaining the factors $1 \ldots p$ with maximal variance to cover 98% of variance and projecting yields low dimensional descriptors $\mathbf{d}_i^{PCA}$. The resulting p-dimensional space $\mathbf{D}^{PCA} = (\mathbf{d}_1^{PCA}, \ldots, \mathbf{d}_N^{PCA})$ thus contains all training data. To obtain a considerable speedup at query time, a kD-tree $\mathcal{K}$ is built from the training data.

Image Query. Given a query or test image $\mathbf{I}_t$ and its associated descriptor $\mathbf{d}_t$, a k-nearest neighbor search is performed in the training space using the kD-tree $\mathcal{K}$, yielding an initial set of m descriptors ($m = 100$ in our experiments). Up to now the euclidean distance assumption was employed and the resulting distances e_{ij} between images i and j are used in the first evaluation approach to estimate the performance contribution of the following step.

To be able to account for rigid image deformations and contrast and brightness variations in the data, the following refined distance metric is employed: For each of the $\mathbf{d}_j \in \mathbf{d}_i, \ldots, \mathbf{d}_m$ miniatures the rotation α, scaling s and translation parameters t_x, t_y are estimated by maximizing the correlation

$$c_j = max \ corr(\mathbf{d}_t, T(\mathbf{d}_j, \alpha, s, t_x, t_y)) \tag{1}$$

where $T(\mathbf{d}_j, \alpha, s, t_x, t_y)$ represents the miniature $\mathbf{d}_j$ after rotation, scaling and translation. The actual maximization is performed using fixed increments for all 4 parameters, which allows for fast optimization using precomputed pixel indices. Using $2 - c_j$ as the new distance measure yields the final order by similarity of the m miniatures for the query image $\mathbf{I}_t$.

While a k-NN approach does not necessarily require a training phase, the kD-tree proves beneficial by ensuring retrieval times in the range of a few milli seconds for the first stage of the retrieval. The optimization of the rigid registration required in the order of a few seconds per query image in a Matlab implementation, which can be easily improved.

3D Volume Retrieval. To perform retrieval of volumetric data, in our case 3D CTs, we extend the approach as follows. Descriptors $\mathbf{d}_i$ are computed as $16 \times 16 \times 16$ volumes and the transformation $T(\cdot)$ now takes into account the additional rotation parameters β and γ as well as the translation t_z. For each tuple $(\mathbf{d}_t, \mathbf{d}_j)$, we optimize

$$c_j = max \ \ corr(\mathbf{d}_t, T(\mathbf{d}_j, \alpha, \beta, \gamma, s, t_x, t_y, t_z)). \tag{2}$$

For computational efficiency, not the entire volume $\mathbf{d}_i$ is transformed, but only the axis-parallel 16×16 slices $\mathbf{d}^x, \mathbf{d}^y, \mathbf{d}^z$ through the center of the volume. This results in the more efficient maximization

$$\mathbf{d}^{xyz} = ((\mathbf{d}^x)^T, (\mathbf{d}^y)^T, (\mathbf{d}^z)^T)^T \tag{3}$$

$$c_j = max \ \ corr(d_t^{xyz}, T(d_j^{xyz}, \alpha, \beta, \gamma, s, t_x, t_y, t_z)). \tag{4}$$

2.2   Distribution Fields (DFs)

The second type of descriptors investigated are Distribution Fields (DFs) [10]. They split an image into b separate channels containing only information from pixels which lie in the corresponding gray level interval. A given normalized image I_i with values in the range $[0, 1]$ is split into a set of channels $\mathcal{C}_i = \{\mathbf{C}_1^i, \ldots, \mathbf{C}_c^i, \ldots, \mathbf{C}_b^i\}$ such that channel $\mathbf{C}_c^i$ at position (x, y) is

$$\hat{\mathbf{C}}_c^i(x, y) = \mathbf{I}_i(x, y) > \frac{c - 1}{b} \wedge \mathbf{I}_i(x, y) < \frac{c}{b}. \tag{5}$$

$$\mathbf{C}_c^i = \hat{\mathbf{C}}_c^i \circ \mathbf{G}(\sigma), \tag{6}$$

where $\mathbf{G}(\sigma)$ performs a spatial smoothing of the channel with a Gaussian filter with standard deviation σ. In our case the images are the miniatures, thus the descriptors are all of the same size. As suggested in [10], we can use the l_1-norm to compute distances d_{ij} between the image descriptors $\mathbf{C}^i$ and $\mathbf{C}^j$:

$$d_{ij} = \sum_{c,x,y} |\mathbf{C}_c^i(x, y) - \mathbf{C}_c^j(x, y)| \tag{7}$$

The miniature volumes employed for the 3D set are treated in the same way, yielding descriptors of size $16 \times 16 \times 16 \times b$. For both the 2D and 3D data set $b = 10$ channels were used.

2.3   Histograms of Gradients (HOGs)

To encode image information through a set of histograms of the gradient orientations in different parts of the image or image patch has seen tremendous success in computer vision [9]. It is also employed in the BVWs methods used for comparison in this paper.

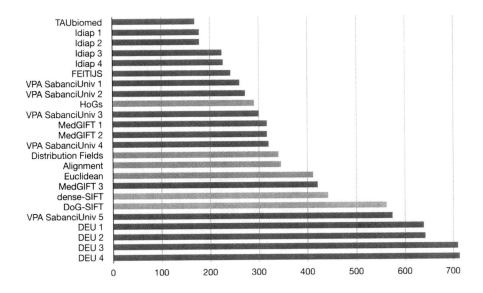

Fig. 2. Error scores for the ImageCLEF 2009 Medical Image Annotation Task [11]. The published benchmark results are depicted in blue, the DOG-SIFT and dense-SIFT approaches in orange and the the methods investigated in this work in green. The error score is based on the evaluation scheme for the IRMA code classification set (Version 2008), with a lower score indicating higher accuracy. Despite its simplicity, the HOG miniatures rank comparatively high, surpassing standard and more optimized BVWs approaches.

Each miniature $\mathbf{I}_i$ is described by one SIFT-like descriptor $\mathbf{H}_i$ as follows. In the 2D case, using the usual 4×4 grid of histograms with 8 bins for the gradient directions $(0, \frac{\pi}{4}, \ldots, \frac{7}{4}\pi)$ yields a 128-dimensional descriptor. In the 3D case $\mathbf{H}_i$ is computed as a $4 \times 4 \times 4$ grid of histograms with $8*8$ directional bins, resulting in a 4096-dimensional descriptor space. The chi-squared distance $\chi^2(\mathbf{H}_i, \mathbf{H}_j)$ is used as distance metric.

3 Experiments

The crucial points to investigate are the retrieval performance of the miniature based descriptors vs. the BVW on different datasets. There are two data sets used for evaluation:

3.1 ImageCLEF Data Set

The ImageCLEF 2009 classification challenge [2] provides a set of 12.677 (training) plus 1.733 (test) 2D radiographs of various body regions. Example images from

---

[2] The ImageCLEFmed 2009 Data Set:
   http://www.irma-project.org/datasets.php?selected=0000900009.dataset

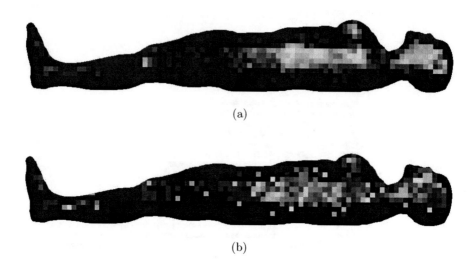

(a)

(b)

Fig. 3. a) Sagittal projection of the distribution of the volumes in the 3D CT data set. For each volume the position of its center in a 3D atlas was annotated as ground truth. b) Mean error between predicted and actual position of the individual volumes using the HOG descriptor, with blue corresponding to 0cm error and red to ≥ 30cm. Note how the accuracy of the prediction strongly correlates with the number of examples available in the database for each location.

this data set can be seen in Fig. 1. All images come labeled with the IRMA code [8], which is a hierarchic multi-dimensional code providing information about modality, body orientation and anatomy.

The task is to annotate each query image with an IRMA code. A kNN search with $k = 3$ was used to find the most similar training examples for each query image. The most often occurring code (or the one of the closest match in parity situations) was assigned to each query image. While the benchmark's rules allow to specify wildcards within the result codes this option was not utilized. Using the provided benchmark script[3] a cumulative error score for each proposed approach (miniature alignment as well as euclidean distance alone, Distribution Fields, Histograms of Gradients) was computed. To provide additional context a standard DOG-SIFT and a dense-SIFT implementation as detailed in [6] were included in the comparison.

3.2   CT Data Set

The second data set is a collection of 3876 3D-CTs extracted from the PACS of the General Hospital Vienna, obtained by querying for all CT data sets of the first days of March 2011. The CTs originate from all different CT scanners present at the department of radiology. For each of the DICOMs the 3D position of its center

---

[3] http://www.idiap.ch/clef2009/evaluation_tools/error_evaluation.eps

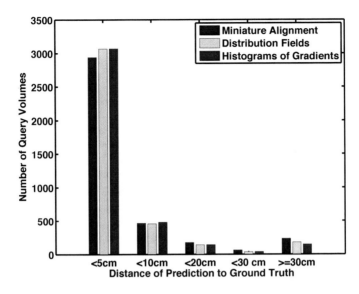

Fig. 4. Histogram of the residual distances in centimeters for the 3D CT dataset. In a leave-one-out setting, the distance of the predicted position to the ground truth is recorded for each query volume.

in a full body atlas was annotated. An overview of the distribution of the centers of the CTs is depicted in Fig. 3a. As can be seen the vast majority of acquisitions are performed with highly standardized protocols, for specific regions where performing CTs is of interest. This means that the image space of these regions is highly populated, whereas there are only 7 CTs of feet in the entire data set.

For a given query image, the task was to predict its position in the reference atlas. The coordinates of the most similar volume (kNN with $k = 1$) were assigned to the each query volume and the euclidean distance in centimeters between resulting position and ground truth were used as error measure. To exclude the influence of misclassification due to left/right similarities (i. e. acquisitions of single hands and feet) all annotations and subsequent evaluations were performed using absolute distances to the sagittal plane for the coordinate orthogonal to the sagittal plane.

4   Results and Discussion

In the following the performance of the proposed approach on the two data sets is detailed.

4.1   ImageCLEF Data Set

Fig. 2 shows the performance comparison of the evaluated methods for the ImageCLEF 2009 data set. The published benchmark results are depicted in

Table 1. Numerical results for the 3D CT data set. While all approaches yield comparable results, HOGs perform best.

Approach	Median	Mean
Miniature Alignment	2.56cm	7.53cm
Distribution Fields (DFs)	2.42cm	6.70cm
Histograms of Gradients (HOGs)	**2.40cm**	**5.72cm**

blue, the implementations of standard DOG-SIFT and dense-SIFT in orange and the methods proposed in this paper in green. The low performance of the standard BVW approaches in comparison to the top result, which also employ BVWs, shows the importance of the careful and data set dependent optimization of the interest point detector, the local descriptors and their integration into visual words. The performance of the published methods (blue) has been carefully optimized on this data set in the last years.

Also of note is, that the reference BVWs (orange) only employ a simple $k = 1$ NN using the χ^2-distance as opposed to more complex classifiers such as hierarchical kernel-SVMs. Also in relation to the miniature based approaches evaluated in this work it has to be kept in mind, that while using elaborated classification schemes increases the classification performance, the ability to yield a ranked list of the most similar training examples gets lost. This would have to be performed, after obtaining the classification result, using the original features (often a combination of several different feature types), using a distance metric which has to be carefully selected.

Some of the image miniature descriptors, without optimizations for this particular data set, perform surprisingly well, given their simple construction and straightforward classification strategy. Retrieving the most similar miniatures using euclidean distance yields the worst results, and rigidly aligning the closest candidates from the kD-tree according to [12] improves the results only slightly. Almost the same performance is achieved using the distribution fields, which mimic the alignment process through the Gaussian smoothing of the individual bins.

Showing the best performance on the 2D data set, and surpassing the established medGIFT framework as well as five BVW approaches is the SIFT-like HOG descriptor, which deals best with the varying contrast and brightness as well as occlusions (implants) and spatial transformations. While it does not outperform more complex and considerably more optimized approaches, it provides an interesting baseline as to how far a simple model can yields useful results with extremely low computational complexity.

4.2   CT Data Set

For the 3D localization task we measured the distance in centimeters between the center of the query image and the center of the most similar image. Fig. 4 shows the histogram of the resulting distances when querying with all images in a leave-one-out fashion. 79.17% of the results for the HOG descriptor are within 5cm of the ground truth, 91.47% are within 10cm. The median residual for the HOG approach is 2.40cm, with the distribution fields and the miniature alignment performing similar at 2.42cm and 2.56cm. Due to the presence of outliers the mean distance are considerably larger: 5.72cm, 6.70cm and 7.53cm, respectively.

The source for the outliers becomes clear when looking at the patterns of their occurrence. Fig. 3 shows the spatial distribution of the mean prediction error for the HOG descriptors. It exhibits a strong correlation between density, i.e. number of images per region, in the data set, and prediction accuracy, with blue areas indicating an error of 0 and red representing $\geq 30cm$. Looking at the lower extremities, for which only few example are available in the data set, the results consist mainly of outliers. On the other hand, in the abdominal, thorax and head region, the most common areas for which CT is performed, localization accuracy is high.

We see this result as confirmation of our hypothesis, that given the highly constrained image space associated with medical images in clinical practice, the use of simple models estimated with the help of large data sets yields promising results and warrants closer investigation. Considering that the results presented in this work are based on only a few days worth of CT acquisitions, we expect the results to significantly improve in areas of lower density by expanding the dataset. We expect a data set with an even spatial distribution, weighted by the expected anatomical and pathological variance, to perform well while limiting the size of the data set.

5   Conclusion and Outlook

In this paper we explored four conceptually simple *holistic* descriptors and metrics derived from image or volume miniatures to perform image retrieval. Results show that the resulting features perform well in medical imaging data retrieval tasks, although they exhibit lower computational complexity compared to, e.g., bags of visual words.

Observations are consistent with the expectation that a dense sampling of the image space is necessary to provide high quality results. Accuracy depends highly on the sampling density provided in the data set, and low performance is limited to the areas of images which rarely occur in the data set. As hypothesized the closed image domain makes the successful application of miniature based approaches already feasible with a few thousand training images compared to many millions in the web image domain. Thus we conclude that the approach of using simple models in conjunction with large data sets presents a promising direction for future research in medical imaging.

Future work will focus on ways to make more disease-specific retrieval feasible including sub-volume retrieval, as well as investigate the robustness of the method on difficult data when expanding the data set to provide a denser sampling in the corresponding areas.

References

1. André, B., Vercauteren, T., Perchant, A., Wallace, M.B., Buchner, A.M., Ayache, N.: Endomicroscopic image retrieval and classification using invariant visual features. In: Proceedings of the Sixth IEEE International Symposium on Biomedical Imaging (ISBI 2009), pp. 346–349. IEEE, Boston (2009)
2. Avni, U., Goldberger, J., Greenspan, H.: Addressing the ImageCLEF 2009 challenge using a patch-based visual words representation. In: Working Notes for the CLEF 2009 Workshop. The Cross-Language Evaluation Forum (CLEF), Corfu, Greece (2009)
3. Dimitrovski, I., Kocev, D., Loskovska, S., Džeroski, S.: ImageCLEF 2009 Medical Image Annotation Task: PCTs for Hierarchical Multi-Label Classification. In: Peters, C., Caputo, B., Gonzalo, J., Jones, G.J.F., Kalpathy-Cramer, J., Müller, H., Tsikrika, T. (eds.) CLEF 2009. LNCS, vol. 6242, pp. 231–238. Springer, Heidelberg (2010)
4. Donner, R., Langs, G., Micusik, B., Bischof, H.: Generalized Sparse MRF Appearance Models. Image and Vision Computing 28(6), 1031–1038 (2010)
5. Feulner, J., Zhou, S.K., Seifert, S., Cavallaro, A., Hornegger, J., Comaniciu, D.: Estimating the body portion of CT volumes by matching histograms of visual words. In: Medical Imaging 2009: Image Processing (Proceedings Volume). vol. 7259, p. 72591V. SPIE (2009)
6. Haas, S., Donner, R., Burner, A., Holzer, M., Langs, G.: SuperPixel-Based Interest Points for Effective Bags of Visual Words Medical Image Retrieval. In: Müller, H., et al. (eds.) MCBR-CDS 2011. LNCS, vol. 7075, pp. 58–68. Springer, Heidelberg (2011)
7. Keysers, D., Dahmen, J., Ney, H., Wein, B., Lehmann, T.: Statistical Framework for Model-Based Image Retrieval in Medical Applications. Journal of Electron Imaging 12(1), 59–68 (2003)
8. Lehmann, T.M., Schubert, H., Keysers, D., Kohnen, M., Wein, B.B.: The IRMA code for unique classification of medical images. In: Medical Imaging 2003: PACS and Integrated Medical Information Systems: Design and Evaluation (Proceedings Volume), vol. 5033, pp. 440–451. SPIE (2003)
9. Lowe, D.G.: Distinctive image features from scale-invariant keypoints. International Journal of Computer Vision 60, 91–110 (2004)
10. Sevilla, L., Learned-Miller, E.: Distribution Fields. Technical Report UM-CS-2011-027, Dept. of Computer Science, University of Massachusetts Amherst (2011)
11. Tommasi, T., Caputo, B., Welter, P., Güld, M.O., Deserno, T.M.: Overview of the CLEF 2009 Medical Image Annotation Track. In: Peters, C., Caputo, B., Gonzalo, J., Jones, G.J.F., Kalpathy-Cramer, J., Müller, H., Tsikrika, T. (eds.) CLEF 2009. LNCS, vol. 6242, pp. 85–93. Springer, Heidelberg (2010)
12. Torralba, A., Fergus, R., Freeman, W.: 80 Million Tiny Images: A large Data Set for Nonparametric Object and Scene Recognition. IEEE Transactions on Pattern Analysis and Machine Intelligence (2008)
13. Ünay, D., Soldea, O., Akyüz, S., Çetin, M., Erçil, A.: Medical image retrieval and automatic annotation: VPA-SABANCI at ImageCLEF 2009. In: Working Notes for the CLEF 2009 Workshop, Corfu, Greece (2009)

Semantic Analysis of 3D Anatomical Medical Images for Sub-image Retrieval

Vikram Venkatraghavan[1] and Sohan Ranjan[2]

[1] Medical Imaging and PACS Lab, School of Medical Science and Technology, Indian Institute of Technology, Kharagpur
[2] Medical Image Analysis Lab, John F. Welch Technology Center, GE Global Research

Abstract. Voluminous medical images are critical assets for clinical decision support systems. Retrieval based on the image content can help the clinician in mining images relevant to the current case from a large database. In this paper we address the problem of retrieving relevant sub-images with similar anatomical structures as that of the query image across modalities. The images in the database are automatically annotated with information regarding body region depicted in the scan and organs present, along with their localizing bounding box. For this purpose, initially a coarse localization of body regions is done in the 2D space taking contextual information into account. Following this, finer localization and verification of organs is done using a novel, computationally efficient fuzzy approximation method for constructing 3D texture signatures of organs of interest. They are then indexed using an inverted-file data structure which helps in ranked retrieval of relevant images. Apart from retrieving sub-images across modalities by image example, automatic annotation and efficient indexing allows query by text, limited only by the semantic vocabulary. The algorithm was tested on a database of non-contrast CT and T1-weighted MR volumes. Quantitative assessment of the proposed algorithm was evaluated using ground-truth database sanitized by medical experts.

Keywords: Semantic analysis, Image retrieval, Organ Localization, CT, MRI.

1   Introduction

Image retrieval plays a central role in medical referencing and patient diagnosis. Digital imaging and communication in medicine (DICOM) tags, used as a retrieval and transfer protocol, encapsulate information regarding the identity of the patient, type of examination, body part examined, anatomical structure and other medical records. DICOM tags contain abstract information and do not scale up to sophisticated data management. Manual annotation of DICOM tags is a cumbersome task and often leads to imprecise image categorization [1]. Content-based image retrieval, which is based purely on low-level features such as color, texture and shape, fails to capture the higher level of abstraction inherent in the medical images. A challenge in developing an efficient medical image retrieval system is to identify semantic regions and anatomical

H. Müller et al. (Eds.): MCBR-CDS 2011, LNCS 7075, pp. 139–151, 2012.

structures for indexing and retrieval. Hence there is a need for an image retrieval algorithm which captures the essence of an image using semantic analysis.

Semantic analysis of anatomical images can be looked at as a multi-organ localization problem. Organ localization is a well-researched topic. Many approaches for organ localization have been proposed for various applications. In the context of image segmentation, organ localization can also be viewed as an approximate initial estimate, which can be further optimized in some sense to obtain a precise segmentation. Many regression-based [2] and classification-based [3], [4] approaches have been proposed to get such an initial estimate. The disadvantage of these approaches is that they are highly modality-specific and in many cases, requires prior knowledge of the organs present in the image in order to localize it. Organ localization can also be done by spatial normalization while registering the patient's image with a pre-constructed atlas. The atlas used can be either a statistical atlas [5], a probabilistic atlas [4] or a multi-organ hierarchical atlas [6]. The disadvantages of using atlas-based methods for organ labeling are: i) the optimal choice of degrees of freedom of the registration model; ii) the optimal choice of reference template and its adaptability for pediatric cases; and iii) adaptability to topological changes in an abnormal condition. Recently, sub-image search based methods for organ localization are being thoroughly investigated. It relies mainly on systematically reducing the search space for each organ within an image and localizing it by minimizing an objective function. S. Seifert et. al. [7] introduced a texture-based algorithm which detects various slices and anatomical landmarks in a CT volume and uses this knowledge to localize six different organs. To the best of our knowledge, this paper is the first to discuss an approach to localize multiple organs in CT as well as MR images using a sub-image search based method.

The selection of a semantic model for annotation and indexing is crucial in semantics-based image retrieval methods. Some of the proposed methods include cross-media relevance model [8] and translation model [9]. In our approach, we use the inverted file model for semantic image retrieval introduced in [10]. An inverted file is a data structure where each term in the semantic vocabulary is associated with related documents in the database along with other necessary information required for ranked retrieval of relevant documents.

Figure 1 shows an overview of the proposed architecture for semantic description and retrieval of images in the database. Given a medical image, initially a coarse localization of various regions is performed and a semantic map is built based on matching with exemplar images. We then introduce a novel fuzzy approximation to develop computationally efficient 3D uniform local Gabor binary pattern texture descriptors for organ detection as well as verification. This information is further used to automatically annotate the images in the database. An inverted-file data structure is used to index the annotated images, which helps in ranked retrieval of images for a query. Automatic annotation in terms of body regions and organs present allows query by image as well as by text, limited only by its semantic vocabulary.

The rest of the paper is organized as follows: In section 2, we describe our proposed algorithm for coarse semantic localization. It is followed by description of the algorithm for validation and fine localization of organs. We also describe our method for creating a semantic index for images in the database. In section 3, experimental results are presented. Finally, we present our conclusions in section 4.

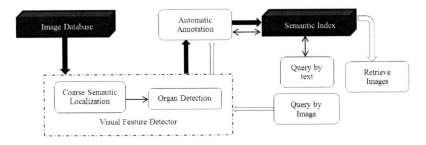

Fig. 1. The proposed system architecture

2   Proposed Approach

2.1   Coarse Semantic Localization

Coarse semantic localization is done by labeling the regions in an input image as one of the following regions: cranial, thoracic, abdominal, sacro-lumbar & the extremities. An approach for estimating body region in CT using SURF descriptors [14] and Hounsfield units is proposed by J. Feulner [12] *et. al.* . The algorithm is not extensible to MR as the absolute gray values in MR images, unlike CT images, are highly dependent on imaging parameters. Moreover, since manual MR examination planning is usually a time-consuming and extremely laborious task [15], minor inconsistencies in alignments between subjects is almost inevitable. Gabor filtering is known to be robust to misalignment in recognition applications [16]. Therefore we have proposed a cascading scheme of multi-scale and multi-orientation Gabor filtering followed by SURF descriptors, which localizes the body portion quite effectively in CT as well as in MR.

We match coronal slices of the given image with that of a pre-selected coronal slice of an exemplar image(s) and select the one which matches it best. Full body CT images are selected as exemplar images for CT, while for MR images, a set of region-specific scans consisting of the aforementioned regions is chosen for exemplar images. Matching is done by computing the Euclidean distance between the SURF descriptors in one Gabor filtered slice with that in the other. After selecting the appropriate coronal slice, we formulate the body region estimation problem as a part-to-whole matching problem. Point-wise correspondences are found for the interest points detected in each slice. The point correspondence minimizes the dissimilarity function defined in equation 1.

$$E = \sum_{i=1}^{N_1} \min_{\forall v_j \in W_i} \left\{ E_{SURF}(u_i, v_j) + \alpha \times E_{Geom}(u_i, v_j) \right\} \tag{1}$$

$U = \{u_i\}_{i=1}^{N_1} \in \Re^2$ are the interest-points in exemplar image. N_1 is the total number of interest points in exemplar image. W_i is a window around the point u_i. Distances between interest points in both images are computed only if they fall within this window. $V = \{v_j\}_{j=1}^{N_2} \in \mathfrak{R}^2$ are interest points in input image within the window W_i. E_{SURF} is the distance between the SURF descriptor values extracted at those points and E_{Geom} is the Euclidean distance between the point coordinates. α is the weighting given to the geometric distance in computing the dissimilarity measure.

Based on the built point correspondences, the body portion depicted in the scan is estimated by propagating the labels from the exemplar image to the image under consideration. An example of point correspondences can be seen in figure 2. Slice-wise processing was chosen over a complete 3D volume processing as the latter is computationally expensive for coarse localizations. As a result, the proposed algorithm is sensitive to large orientation changes typically observed in MR scans of a similar region using different protocols. The algorithm does not intrinsically take care of it. It is taken care of by providing relevant training image(s).

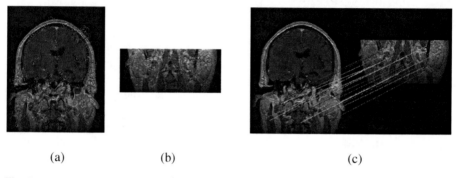

    (a)   (b)   (c)

Fig. 2. (a) Represents the interest points detected in an exemplar T1-weighted MR image. (b) Shows matched coronal slice of the input 3D volume, along with the interest points detected. (c) The point correspondence is shown by connecting the 2 interest points. The image is best viewed in color.

2.2   Organ Detection

After a coarse localization is done, fine 3D localization of organs and verification is done using volume-based texture descriptors. We use Local Gabor Binary Pattern (LGBP) for this purpose. We also introduce a computationally efficient way to compute fuzzy histogram of uniform LGBP. For this purpose, fuzzy uniformity index (FUI), a fuzzy measure for calculating the uniformity of a pixel in its neighborhood, is introduced. Apart from this, histogram of FUI itself acts as one of the discriminating features for organ localization. The following sub-sections describe FUI and LGBP in detail as well describe the methodology used for localizing organs in 3D.

Fuzzy Uniformity Index. For the computation of Uniform Local Binary Pattern (ULBP) [21], Ojala *et. al.* classified each pixel as either uniform or non-uniform. The methodology used by them for that purpose is given below:

A circular neighborhood is considered around a pixel. 'P' points are chosen on the circumference of the circle such that they are all equidistant from the central pixel.

Uniformity is measured as a function of number of bit transitions in the circular bit stream of these 'P' points, obtained by using the center pixel's gray value as threshold. A look-up table is generally used to compute the bit-transitions in order to reduce computational complexity.

Classification of a pixel into the above two classes in 3D can be done by either creating a look-up table or by connected component labeling (CCL) algorithm in the resultant binary sub-image. Computing CCL for neighborhood around each voxel separately is a computationally expensive step. A look-up table, even for the smallest 3D neighborhood would have 2^{26} elements, which is again impractical to implement. Hence to overcome this disadvantage, a fuzzy classification approach is proposed here. In this approach, every pixel is classified as both uniform as well as non-uniform, but to varying degrees. The classification is done as follows: Let degree of uniformity of a pixel in the neighborhood be FUI_p. Then,

$$FUI_p = f(|g_p - g_c|) \tag{2}$$

where f(x) is a monotonically decreasing fuzzy membership function for x > 0 and f(0) = 1. This degree of uniformity is termed as Fuzzy Uniformity Index. g_c is the gray value of the center voxel and g_p is the gray value of its neighbors. Median of the FUI_p values in the neighborhood is denoted as the fuzzy uniformity index of the center voxel. Histogram of FUI is taken as one of the discriminating features for organ detection. The FUI histograms generated by various membership functions were ranked based on Feature Usability Index [23], a classifier-independent feature ranking technique, and the results are reported in table 1. Feature Usability Index for each bin is computed for each bin separately and histogram usability index (HUI) is calculated as a mean of those values.

Table 1. Ranking of FUI histograms

Membership Function Parameters		Histogram Usability Index	Rank
Gaussian (μ, σ)	(0, 25)	3762.54	4
	(0, 50)	7425.71	1
	(0, 15)	5325.33	2
Triangular (m)	(15)	1871.12	7
	(30)	2321.36	5
	(45)	4336.64	3
	(60)	2074.23	6

μ and σ are the mean and standard deviation of the Gaussian membership function. Triangular membership function is selected such that the membership values decrease linearly between 0 and m and is 0 beyond m.

3D Local Gabor Binary Pattern. 3D LGBP, which is a combination of the Gabor wavelets and the LBP operator, can be defined in 3D as

$$LGBP_{P,R}^{NU} = \sum_{p=0}^{P-1} s(G_p - G_c) \tag{3}$$

where G_c is the magnitude of 3D Gabor filtered image at the central voxel, G_p is the magnitude of its neighbors, and s(.) is a Heaviside step function..The superscript NU for LGBP is given in the equation to denote that the LGBP defined here is different from that of traditional LGBP [17] and Gabor-cascaded version of 3D uniform LBP [11, 22]. 3D Gabor Filter is defined as

$$h(x, y, z) = g(x', y', z') \times s'(x, y, z) \qquad (4)$$

where $g(x', y', z')$ is a 3D Gaussian envelope, $s'(x, y, z)$ is a complex sinusoidal function and (x', y', z') are the rotated spatial co-ordinates of the Gaussian envelope. For a detailed set of equations and other explanations, refer [20]. A fuzzy histogram of $LGBP_{P,R}^{NU}$ is constructed by increasing the bins by FUI value at each voxel. Various Local Gabor Binary Pattern Fuzzy Histogram Sequence (LGBPFHS) were constructed by varying parameters for each organ under consideration. Ranking of these features was done based on Feature Usability Index [23], the results of which are presented in table 2.

Table 2. Feature Ranking of LGBPFHS

Parameters					Value	Rank
λ	σ_x	σ_z	θ	φ	HUI	
						3
5.88	1.8	0.6	0	45	8290.93	
3.92	1.8	0.3	0	45	46998.39	1
5.88	0.9	0.3	0	45	2519.55	5
3.92	0.9	0.2	0	45	1418.74	7
5.88	1.8	0.6	45	90	1414.48	8
3.92	1.8	0.3	45	90	7571.83	
5.88	0.9	0.3	45	90	1480.20	4
3.92	0.9	0.2	45	90	20885.21	6
						2

λ denotes the wavelength of the complex sinusoidal envelope of the Gabor filter in mm. σ_x denotes the standard deviation of the 3D Gaussian curve in the x- (as well as y-) axes. σ_z denotes the standard deviation of the 3D Gaussian curve in the z- axis. θ and φ are the orientations of the Gabor filter.

Sliding Window Search. Histogram of FUI and fuzzy histogram of LGBP are used as discriminating features for organ detection. Coarse semantic localization, discussed in the previous section, reduces the search space for smaller organs. For a complete 3D localization of organs, we use a sliding window search in the reduced search space. The histograms with the best rank (in table 1 and 2) were considered as model histograms for the search phase (Figure 3). Each image in the database was divided into a series of overlapping windows. Selection of window size is a crucial step for detection based on sliding window. We use a data-driven approach for selection of an appropriate window size. We use the previously described Gabor- SURF cascading scheme to detect the starting and ending axial slices of the organ(s) of interest. This not only estimates the organ size in z-direction, but also reduces the search space for detection even further. The size of the organ is approximated in X- and Y- directions based on the aspect-ratio information obtained from training data. Features computed for the window are compared with the model histograms using chi-squared distance measure [13].

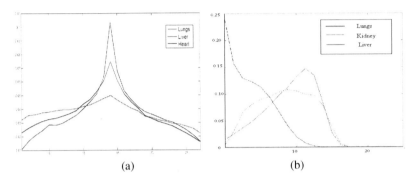

(a)   (b)

Fig. 3. Depicts mean model histograms in CT (a) Model LGBPFHS histogram (b) FUI histogram

2.3   Inverted File Generation

The organs thus localized from CT and MR images in the database are indexed using an inverted-file data structure [10]. For this purpose, a semantic vocabulary is formed based on the organs detected and body regions localized in the database. The semantic vocabulary S consists of:

$$S = \{Liver, Lung(R), Lung(L), Heart, Kidney(R), Kidney(L), Spleen, Head, Neck,$$

$$Abdomen, Thorax, Cranium, Sacro - lumbar \ \& \ extremities\}$$

An inverted file is a data structure where each term in the semantic vocabulary is associated with related documents in the database along with other necessary information. For each term in the semantic vocabulary ($term_j$), the following information is collected in the inverted file: i) Image frequency $\{IF_j\}$: the number of images containing $term_j$. ii) Percentage of volume occupied $\{vo_j(i)\}$: the percentage of volume a structure representing $term_j$ occupies in the image i *iii)*Image ID $\{doc_j\}$: the documents that contain $term_j$ *iv)* Localizing Bounding Box $\{BB_j(i)\}$: the Coordinates of the localizing bounding box for the structure in image i for term j.

Weighting for $term_j$ on $Image_i$ is inversely proportional to IF_j and directly proportional to $vo_j(i)$.

$$W(i, j) = \log(\frac{N}{IF_j}) \times vo_j(i) \tag{5}$$

The relevance of an image in the database to a query (term or image) is calculated by computing $W(i, j)$ for all the images in the database and ranking them.

3   Experimental Results

In this section, we illustrate the results of various steps explained in previous sections as well as compare various existing methods.

Coarse semantic localization for a CT image in the database is illustrated in figure 4. Fig. 4(a) & (b) show a coronal slice of an exemplar image and image in the database respectively. The lines connecting various points in the 2 images represent the mapping between corresponding points.

Figure 5 shows the variation of dissimilarity function for detecting the slice where kidney starts in MR, for SURF, SIFT [18], GIST [19] and our proposed scheme. Table 3 compares the mean and standard deviation values of Hessian at the minima across all retrievals performed. It can be inferred from Table 3 that the proposed cascading scheme has greater mean value of hessian at the minima and thus serves as a stronger objective function. This can also be verified from fig. 5, as our proposed scheme has a stronger minimum when compared with traditional methods.

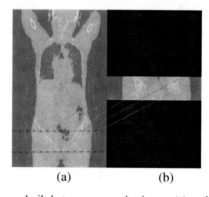

(a)   (b)

Fig. 4. Point Correspondence built between exemplar image (a) and an image in the database (b) is shown. The body portion estimated is shown as a dotted line in 4(a). The corresponding label is annotated for the input-image.

Table 3. Hessian at minima

Method	Hessian at minima	
	Mean	Std. Dev.
SURF	8.6356	3.5712
SIFT	7.0264	3.0478
GIST	4.9123	2.1227
Gabor + SURF	13.1546	4.2639

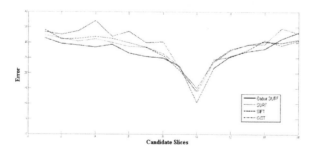

Fig. 5. Variation of the objective function used for axial slice matching is shown for 4 different methods here

Table 4. Comparison of Time required.

Size of the volume	Fuzzy Histogram of 3D LGBP	Traditional 3D Uniform LGBP
9×9×9	1.2 ms	1.7357 s
16×16×16	6.7 ms	9.7524 s
25×25×25	25.5 ms	37.2024 s

Table 5. Comparison with other algorithms: Bounding box localization errors (Mean, Std. Dev. in mm)

Organs	Head	Kidneys		Liver	Lungs		Heart	Spleen
		Right	Left		Right	Left		
Proposed	17.7	19.4	18.1	18.9	15.3	16.8	17.3	18.7
Algorithm	9.6	14.3	15.7	14.2	9.9	10.2	14.2	19.5
Regression	-	18.5	17.3	17.1	15.6	17.0	15.4	20.7
Forests [2]		18.0	16.5	16.5	16.3	17.2	15.5	22.8
Decision	29.9	44.5	25.4	22.68	26.75	27.05	21.3	-
Forests [3]	12.8	15.0	9.8	5.3	9.44	7.25	5.7	

As our purpose for introducing the fuzzy approximation in the computation of local Gabor binary pattern here is to reduce the computational complexity of constructing a texture descriptor using 3D Uniform LGBP, we only compare the time required to compute the two descriptors over various 3D volumes. The comparison of the texture description capability over standard texture databases is beyond the scope of this paper. Both the algorithms were implemented using ITK library and were run on an Intel Core 2 Duo CPU with a processor speed of 2.6 GHz. The result is reported in Table 4.

Localization accuracy is assessed for the proposed algorithm by using repeated random sub-sampling cross validation technique. The location of each organ center is computed and compared with available ground truth images. The results are reported in Table 5. It is also compared with the prevailing state-of-the-art organ localization techniques for CT. The results from [2] and [3] are reported directly from their respective papers. The proposed algorithm localizes organs in CT as well as in MR, while the reported results in [2] and [3] are for CT alone. The fields that are marked '-' are the ones which were not reported in the corresponding paper. Table 5 clearly indicates that our proposed algorithm outperforms the decision-forest based algorithm proposed in [3] in terms of mean and standard deviation of the localization errors. It can also be seen that our method is better for smaller organs when compared with [2]. While the error for larger organs are comparable but slightly greater than that obtained by regression forests, its adaptability to multiple modalities makes it better than [2] as well.

For the purpose of retrieval, the algorithm is tested on a database consisting of 80 T1-weighted MR volumes and 60 non-contrast CT volumes. Figure 6 shows a sample query image and corresponding retrieved images. The top ranked image for a query CT image is an MR image which captures similar regions. It can also be noted from figure 6 that as the captured region in the image increases, the relevance for a query decreases, as the percentage of abdomen region as well as other relevant organs decreases. Figure 7 shows the gradual decrease of weight (W) with respect to the rank of retrieval.

Images can also be retrieved using query text. Apart from the semantic vocabulary, the queries can also contain keywords such as 'partial' or 'full', which will look only for images with partial or full organs respectively. The keyword may also specify the modality of the image. Figure 8 shows top ranked hit for a sample query by text.

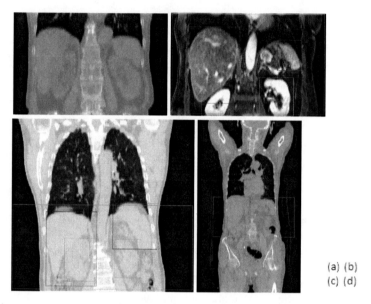

(a) (b)
(c) (d)

Fig. 5. (a) Query Image: a CT volume (b-d) Retrieved images: Rank 1, 10, and 20 respectively, with relevant parts marked. Bounding box for liver is Liver is red, Kidneys are green, Spleen is pink and Abdomen is orange. Only 1 coronal slice is shown for each volume. The weight (W) obtained for the images (b-d) are: 0.5920, 0.2679 and 0.1276 respectively.

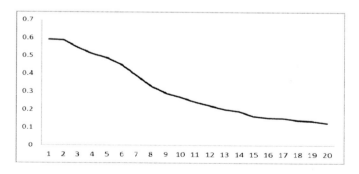

Fig. 6. Variation of weight (W) with respect to rank of retrieval

Fig. 7. Retrieved image (Rank 1) with relevant organs localized for query by text: 'Head', 'Neck', 'MR'

4   Conclusion

We have introduced an approach to retrieve related images from a multi-modal and semantically indexed database of medical images. The major contributions in this context are: the development of a framework for creating a semantic index of 3D anatomical images based on coarse semantic regions and organs, which allows retrieval of relevant anatomical structures across modalities; the development of a fuzzy approximation technique which decreases the time required for construction of histogram of local Gabor binary pattern by the order of 10^3; and the development of a multi-organ localization methodology which localizes organs in CT as well as MR volumes without using atlas information.

The prototype is currently being tested on non-contrast CT and T1-weighted MR volumes. The algorithm for semantic classification and annotation generation does not require any prior knowledge on the existence of various organs nor does it assume the existence of various supporting structures or slices. This makes it ideal for retrieval applications where large variations are seen in the images for the database. The key issue for semantic classification by this approach is the identification of a texture measure(s) which distinguishes various regions and organs. The fuzzy approximation for the construction of local Gabor binary pattern histograms and FUI histograms helps in developing a computationally efficient way to generate unique organ signatures. The inverted-file model for semantic indices helps in ranked retrieval of relevant images from the database. Finally, we would like to extend our work to include MRI images of other protocols as well expand the semantic vocabulary by adding smaller organs to it, by exploiting the hierarchical nature of inter-organ relationships.

Acknowledgement. The authors would like to thank Dr. Jyotirmoy Banerjee, GE Global Research, Prof. Ajoy Ray, IIT Kharagpur and Dr. Jyotirmoy Chatterjee, IIT Kharagpur, for their valuable assistance in this research, as also Dr. Jerome Knoplioch, GE Healthcare, for providing the data necessary for carrying out this research.

References

[1] Gueld, M.O., Kohnen, M., Keysers, D., Schubert, H., Wein, B.B., Bredno, J., Lehmann, T.M.: Quality of DICOM headers information for image categorization. In: Proceedings of SPIE Medical Imaging, vol. 4685, pp. 280–287 (2002)

[2] Criminisi, A., Shotton, J., Robertson, D., Konukoglu, E.: Regression Forests for Efficient Anatomy Detection and Localization in CT Studies. In: Menze, B., Langs, G., Tu, Z., Criminisi, A. (eds.) MICCAI 2010. LNCS, vol. 6533, pp. 106–117. Springer, Heidelberg (2011)

[3] Criminisi, A., Shotton, J., Bucciarelli, S.: Decision forests with long-range spatial context for organ localization in CT volumes. In: MICCAI Workshop on Probabilistic Models for Medical Image Analysis, MICCAI-PMMIA (2009)

[4] Hyunjin, P., Bland, P.H., Meyer, C.R.: Construction of an abdominal probabilistic atlas and its application in segmentation. IEEE Transactions on Medical Imaging 22(4), 483–492 (2003)

[5] Yoshida, Y., Chen, Y.W., Okada, T., Yokota, F., Sato, Y., Hori, M.: Representation and evaluation of statistical prediction powers of neighboring organ shapes for construction of multi-organ statistical atlas. In: 2nd International Conference on Software Engineering and Data Mining (SEDM), pp. 696–699 (2010)

[6] Okada, T., Yokota, K., Hori, M., Nakamoto, M., Nakamura, H., Sato, Y.: Construction of Hierarchical Multi-Organ Statistical Atlases and Their Application to Multi-Organ Segmentation from CT Images. In: Metaxas, D., Axel, L., Fichtinger, G., Székely, G. (eds.) MICCAI 2008, Part I. LNCS, vol. 5241, pp. 502–509. Springer, Heidelberg (2008)

[7] Seifert, S., Barbu, A., Zhou, S.K., Liu, D., Feulner, J., Huber, M., Suehling, M., Cavallaro, A., Comaniciu, D.: Hierarchical parsing and semantic navigation of full body CT data. In: Proceedings of SPIE, vol. 7259 (2009)

[8] Jeon, J., Lavrenko, V., Manmatha, R.: Automatic image annotation and retrieval using cross-media relevance models. In: Proc. of the 26th Annual International ACM SIGIR Conference on Research and Development in Information Retrieval, pp. 119–126. ACM, New York (2003)

[9] Duygulu, P., Barnard, K., de Freitas, J.F.G., Forsyth, D.: Object Recognition as Machine Translation: Learning a Lexicon for a Fixed Image Vocabulary. In: Heyden, A., Sparr, G., Nielsen, M., Johansen, P. (eds.) ECCV 2002. LNCS, vol. 2353, pp. 97–112. Springer, Heidelberg (2002)

[10] Zhang, D., Islam, M.M., Lu, G., Hou, J.: Semantic Image Retrieval Using Region Based Inverted File. In: Proc. of Digital Image Computing: Techniques and Applications, pp. 242–249 (2009)

[11] Fehr, J., Burkhardt, H.: 3D rotation invariant local binary patterns. In: Proc. of 19th International Conference on Pattern Recognition, pp. 1–4 (2008)

[12] Feulner, J., Zhou, S.K., Seifert, S., Cavallaro, A., Hornegger, J., Comaniciu, D.: Estimating the body portion of CT volumes by matching histograms of visual words. In: Proc. of SPIE Medical Imaging, Lake Buena Vista, Florida (2009)

[13] Huong, V.T.L., Park, D.C., Woo, D.M., Lee, Y.: Centroid neural network with Chi square distance measure for texture classification. In: Proc. of International Joint Conference on Neural Networks, pp. 1310–1315 (2009)

[14] Bay, H., Ess, A., Tuytelaars, T., Gool, L.V.: SURF: Speeded Up Robust Features. Computer Vision and Image Understanding (CVIU) 110(3), 346–359 (2008)

[15] Peng, Z., Zhong, J., Wee, W., Lee, J.: Automated Vertebra Detection and Segmentation from the Whole Spine MR Images. In: Proc. of 27th Annual International Conference of the Engineering in Medicine and Biology Society, pp. 2527–2530 (2006)

[16] Shan, S., Gao, W., Chang, Y., Cao, B., Yang, P.: Review the strength of Gabor features for face recognition from the angle of its robustness to misalignment. In: Proc. of International Conference on Pattern Recognition, pp. 338–341 (2004)

[17] Zhang, W., Shan, S., Gao, W., Zhang, H.: Local Gabor Binary Pattern Histogram Sequence (LGBPHS): A novel non-statistical model for face representation and recognition. In: Proc. Of International Conference on Computer Vision, Beijing, China, pp. 786–791 (2005)

[18] Lowe, D.: Distinctive image features from scale-invariant keypoints. International Journal of Computer Vision 60(2), 91–110 (2004)

[19] Oliva, A.: Gist of the scene. Neurobiology of Attention, pp. 251–256 (2005)

[20] Qian, Z., Metaxas, D.N., Axel, L.: Extraction and Tracking of MRI Tagging Sheets Using a 3D Gabor Filter Bank. In: Proc. of 28th Annual International Conference of the IEEE on Engineering in Medicine and Biology Society, pp. 711–714 (2006)

[21] Ojala, T., Pietikainen, M., Maenpaa, T.: Multiresolution gray-scale and rotation invariant texture classification with local binary patterns. IEEE Transactions on Pattern Analysis and Machine Intelligence 24(7), 971–987 (2002)

[22] Paulhac, L., Makris, P., Ramel, J.Y.: Comparison between 2D and 3D Local Binary Pattern Methods for Characterisation of Three-Dimensional Textures. In: Proceedings of the 5th International Conference on Image Analysis and Recognition, pp. 670–679 (2008)

[23] Sheet, D., Ray, A.K., Chatterjee, J.: Feature Usability Index. LAP Lambert Academic Publishing (2011)

Author Index